Springer Proceedings in Materials

Volume 19

Springer Proceedings in Materials publishes the latest research in Materials Science and Engineering presented at high standard academic conferences and scientific meetings. It provides a platform for researchers, professionals and students to present their scientific findings and stay up-to-date with the development in Materials Science and Engineering. The scope is multidisciplinary and ranges from fundamental to applied research, including, but not limited to:

- Structural Materials
- Metallic Materials
- Magnetic, Optical and Electronic Materials
- Ceramics, Glass, Composites, Natural Materials
- Biomaterials
- Nanotechnology
- Characterization and Evaluation of Materials
- Energy Materials
- Materials Processing

To submit a proposal or request further information, please contact one of our Springer Publishing Editors according to your affiliation:

European countries: **Mayra Castro** (mayra.castro@springer.com)
India, South Asia and Middle East: **Priya Vyas** (priya.vyas@springer.com)
South Korea: **Smith Chae** (smith.chae@springer.com)
Southeast Asia, Australia and New Zealand: **Ramesh Nath Premnath** (ramesh.premnath@springer.com)
The Americas: **Michael Luby** (michael.luby@springer.com)
China and all the other countries or regions: **Mengchu Huang** (mengchu.huang@springer.com)

This book series is indexed in **SCOPUS** database.

Mohammad Jawaid · Mohamad Midani ·
Ramzi Khiari
Editors

Proceedings of 2nd World Conference on Byproducts of Palms and Their Applications

ByPalma 2021, Kuala Lumpur, Malaysia

Springer

Editors
Mohammad Jawaid
Universiti Putra Malaysia
Serdang, Malaysia

Mohamad Midani
German University in Cairo
Cairo, Egypt

Ramzi Khiari
Higher Institute of Technological Studies
in Ksar-Hellal
Ksar Hellal, Tunisia

ISSN 2662-3161 ISSN 2662-317X (electronic)
Springer Proceedings in Materials
ISBN 978-981-19-6197-7 ISBN 978-981-19-6195-3 (eBook)
https://doi.org/10.1007/978-981-19-6195-3

This Springer imprint is published by the registered company Springer Nature Singapore Pte Ltd.
The registered company address is: 152 Beach Road, #21-01/04 Gateway East, Singapore 189721, Singapore

Preface

The palm plantations are common components of natural and cultivated flora around the world, including a wide variety of species (e.g., date palm, coconut, oil palm, sugar palm, doum palm, etc.). These palm plantations are the main source of livelihood for significant sectors of the world population especially in the south. The main product of palm plantations (cash crop) was the focus of the scientific and industrial community worldwide, whereas their byproducts (secondary products) received little attention. Despite that, palms generate huge amount of byproduct of pruning and fruit processing, which represent a burden on the growers and processors, and may cause fire accidents and infestation by dangerous insects. From a sustainable development perspective, palm byproducts are considered additional renewable resources obtained from the same investment made to cultivate the primary product (fruits), such as, land, water, labor, fertilizers, and pesticides. These byproducts may represent a sustainable material base for a wide spectrum of industries ranging from compost, medium density fiber boards (MDF), block boards, and pulp, up to fiber reinforcements for advanced composites. Thus, there is a need to rediscover palm byproducts and maximize their added value via industrial technological advancement that can help in the sustainable development of vast rural areas around the world.

Proceedings of 2nd World Conference on Byproducts of Palms and Their Applications ByPalma 2021, Kuala Lumpur, Malaysia highlight the great potential of the palm byproducts in the circular bioeconomy of the future! The book is divided into four parts on the use of palm byproducts in wood and boards, polymeric composites, cementitious composites, and other applications.

We are highly thankful to all contributing authors who provided their valuable ideas and knowledge in this edited book. We attempt to gather all the scattered information of authors from diverse fields around the world in the area of palm byproducts utilization and finally complete this venture in a fruitful way.

We are highly thankful to Springer Nature team for their generous cooperation at every stage of the book production.

Serdang, Malaysia — Mohammad Jawaid
Cairo, Egypt — Mohamad Midani
Ksar Hellal, Tunisia — Ramzi Khiari

Contents

About the Editors

Dr. Mohammad Jawaid is currently working as Senior Fellow (Professor) at Biocomposite Technology Laboratory, Institute of Tropical Forestry and Forest Products (INTROP), Universiti Putra Malaysia (UPM), Serdang, Selangor, Malaysia, and also has been Visiting Professor at the Department of Chemical Engineering, College of Engineering, King Saud University, Riyadh, Saudi Arabia, since June 2013. He has more than 20 years of experience in teaching, research, and industries. His area of research interests includes hybrid composites, lignocellulosic reinforced/filled polymer composites, advance materials: graphene/nanoclay/fire retardant, modification and treatment of lignocellulosic fibers and solid wood, biopolymers and biopolymers for packaging applications, nanocomposites and nanocellulose fibers, and polymer blends. So far, he has published 55 books, 75 book chapters, more than 400 peer-reviewed international journal papers, and several published review papers under top highly cited articles in ScienceDirect. He also obtained five Patents and six Copyrights. H-index and citation in Scopus are 68 and 21578, and in Google scholar, H-index and citation are 82 and 29647. He worked as Chief Executive Editor of Pertanika UPM Journals from 2021–2022. He is founding Series Editor of Springer-3 Series: Composite Science and Technology; Sustainable Materials and Technology; and Smart Nanomaterials and Technology and also Series Editor of Springer Proceedings in Materials, Springer Nature and also International Advisory Board Member of Springer Series on Polymer and Composite Materials. He worked as Guest Editor of special issues of SN Applied Science, Frontiers in Sustainable Food Systems, Current Organic Synthesis and Current Analytical Chemistry, International Journal of Polymer Science, IOP Conference Proceeding. He is also Editorial Board Member of Journal of Polymers and The Environment, Journal of Natural Fibres, Journal of Plastics Technology, Applied Science and Engineering Progress Journal, Journal of Asian Science, Technology and Innovation, and the Recent Innovations in Chemical Engineering. Besides that, he is also Reviewer of several high-impact international peer-reviewed journals of Elsevier, Springer, Wiley, Saga, ACS, RSC, Frontiers, etc.

Dr. Mohamad Midani is Assistant Professor at the Materials Engineering Department, German University Cairo, and Adjunct Assistant Professor at Wilson College of Textiles, North Carolina State University. He is also Managing Partner of inTEXive consulting and Secretary General of the International Association of Palm Byproducts. Dr. Midani has long experience in the textile industry, with a focus on technical textiles and composites, with an ongoing research collaboration and consulting to the industry. His research mostly focuses on natural fibers and their composites, and he is the Co-Inventor of the long textile fibers from date palm byproducts (PalmFil). He is also Co-Founder of the World Conference on Palm Byproducts and their Applications (ByPalma). He published more than 30 articles in international journals and conferences in addition to three books. Dr. Midani is Reviewer of several international journals from different publishers. He teaches undergraduate and graduate courses on Fiber Reinforced Polymer Composites as well as New Product development and Innovation Management. In 2022, Dr. Midani received the distinguished young alumni award from Wilson College of Textiles NC State University in recognition for his impact in the textile field. Dr. Midani has his B.Sc. in Mechanical Engineering from Ain Shams University (2006), M.T. in Textiles & Apparel Technology and Management (2012), and Ph.D. in Fiber and Polymer Science from the College of Textiles NC State University (2016). Dr. Midani is certified New Product Development Professional (NPDP) and Member of the Product Development and Management Association (PDMA) and Co-Founder and Member of the board of PDMA Egypt chapter.

Dr. Ramzi Khiari is Associate Professor at Institut Supérieur des études Technologiques de Ksar-Hellal (Monastir, Tunisia) in the department of Textile Engineering and Research and Development Specialist at Laboratory of Pulp and Paper Science and Graphic Arts LGP2 in Grenoble Institute of Technology, University of Grenoble. Dr. Ramzi has an H-index of 26 and co-authored more than 80 publications receiving 2353 citations. His research interests focus on the valorization of biomass at multi-scale levels, namely: fibers, nanocellulose, lignin, hemicelluloses, and their use as potential raw material in several industrial applications (textile, papermaking, polymeric materials, composites, and nanocomposites). A particular focus is given for vegetal biomass from annual plants and particularly agricultural residues and industrial wastes.

Palm Byproducts Wood and Boards

Physical and Mechanical Properties of Wood from Date Palms Related to Structure

Leila Fathi, Arno Frühwald, and Mohsen Bahmani

Abstract Date palms belong to the family of *Arecaceae* with several species. In the current study, density and compression strength of the wood from two date palm trunks (*Phoenix dactylifera*) and their relation to wood structure as well as strength properties of vascular bundles were investigated. Results showed that—different from most other palms—there is no significant difference in density across the trunk diameter and along the trunk length. Wood compression strength (MOR) parallel to fiber direction correlates significantly with wood density, but it does not correlate with a share of vascular bundles on the sample cross-section. The reason is the weak parenchyma which allows buckling of the vascular bundles and shear failures along the interface parenchyma and vascular bundle. MOE and MOR of single vascular bundles are much higher compared to the overall wood. But this reinforcement has its limits due to the much weaker parenchyma. The results contribute to developing a mechanical model to describe material characteristics based on the structure and density of the palm wood.

Keywords Palms · Wood density · Volume · Utilization

1 Introduction

Palm trees are a family of plants (*Arecaceae*). Most of them are trees, but some are shrubs. Most species grow in tropical regions, but some in subtropical or semi-arid regions like date palms [1–4]. Economically important palm species are coconut

L. Fathi (✉) · M. Bahmani
Department of Natural Resources and Earth Sciences, Shahrekord University, Shahrekord, Iran
e-mail: leila.fathi@sku.ac.ir

M. Bahmani
e-mail: mohsen.bahmani@sku.ac.ir

A. Frühwald
Section Mechanical Wood Technology, Centre of Wood Science, University of Hamburg, Hamburg, Germany
e-mail: arno.fruehwald@uni-hamburg.de

M. Jawaid et al. (eds.), *Proceedings of 2nd World Conference on Byproducts of Palms and Their Applications*, Springer Proceedings in Materials 19,
https://doi.org/10.1007/978-981-19-6195-3_1

palm, oil palm, date palm. Palms are one of the widely planted tree species. They have been important to humans throughout the history. Palms are used for food, animal feed, and many other everyday products and are commonly found in parks and gardens. A palm tree is divided into roots, trunk and crown. The structure of palm trunks is simple compared to most trees. Palms do not have a bark. Instead, the epidermis hardens to form a protective layer. The trunk consists of a ground parenchymatous tissue and embedded vascular bundles. One terminal growing point (apical meristem) is protected underneath a layer of leaves. Once the palm nearly reaches final thickness, it begins to grow in height. Growth can be very slow, lasting several years because the apical bud must reach a certain size before the trunk can develop [5]. Due to the absence of a cambium, palm trees have no secondary growth. The increase in diameter is limited and caused by cell division and cell enlargement in the parenchymatous ground tissue as well as in the enlargement of vascular bundles [2, 5].

A slightly different situation is related to date palm as one of the important palm species. Date palms grow in semi-arid areas (a total of about 80 Mio palms which means some close to 1000,000 ha), mainly in the Middle East and Northern Africa, where timber shortages happen even throughout history, and the gap between timber demand and supply is rather big. Overall it is assumed that palms will play an increasing role in natural fibers supply for the wood sector in total. But palm trunk fibers are very much different from "normal wood fibers". Palm trunks, leaves, and fruit stands are the main fiber sources of which the trunk is closest to "normal wood" in its composition, structure, and shape of the trunk. Total availability of 200 mio m^3/year or more of trunk volume (~75% oil palm, 20% coconut palm, and 5% date palm) challenges to research and development for utilizing this material. Palms are monocotyledons which means that the structure of the trunk and therefore the "wood" is different from "normal/common timber". This is mainly relevant for the tissue structure and composition, density, water content, ash and silica content, and sugar/starch content. These features are much relevant for palm wood processing, product design, and product properties/performance. One basic area of concern for wood utilization is the physical/mechanical properties of the material. Limited systematic and comprehensive research exists. Therefore, this study aims to partially bridge the knowledge gap regarding the structure and physicomechanical properties of date palm wood.

2 Materials and Methods

Two date palm trees (*Phoenix dactylifera*) were felled with 8 m height from a plantation in Khuzestan Province, the Ahvaz countryside in the southwest Iran.

The date palm trunks were about 8 m high up to the beginning of the leaves. The trunks were cut with a chainsaw into 6 disks of 10–15 cm thickness from different trunk heights (8 m). The disks were further cut into three sections. In order to avoid fungal colonization, the samples were stored directly in a deep freezer (within 12 h) after sawing. From each disk, sections were used for physical and mechanical properties testing. Physical properties of palm wood were investigated, including the wood density. The density determination is based on DIN 52,183 [6] and is defined as the relationship of mass to volume. The wood density of palm wood was determined on the basis of air-dried specimens (mc ~ 8–10%). The following mechanical properties of palm wood were investigated: tensile strength of single vascular bundles and wood compression strength parallel to the grain.

3 Results and Discussion

An important factor in determining the mechanical properties of coconut and oil palm wood is the density of wood as it increases from the center of the trunk towards the "bark". In date palm, however, there is no significant difference across the trunk. On the other hand, it is proved that at the bottom of the trunk, the density of coconut and oil palm wood is closely related to the distribution of VBs ($R^2 = 0.995/R^2 = 0.97$). However, Fig. 1 shows no relationship between density and the distribution of VBs ($R^2 = 0.03$) in the date palm wood.

The density of date palm wood, which is closely related to the densities of VBs and ground tissue, does not have a distinct variation between three different zones along the trunk diameter as well as along the trunk height. It can be interpreted that, for grading lumber and for developing sawing patterns in sawmill processing, other parameters are also effective; mainly the parenchyma tissue, chemical composition of wood, i.e., lignin, cellulose and hemicellulose and the other anatomical parameters, i.e., proportion of fibers in a vascular bundle, cell wall thickness, and microfibrillar angle in the cell wall layers.

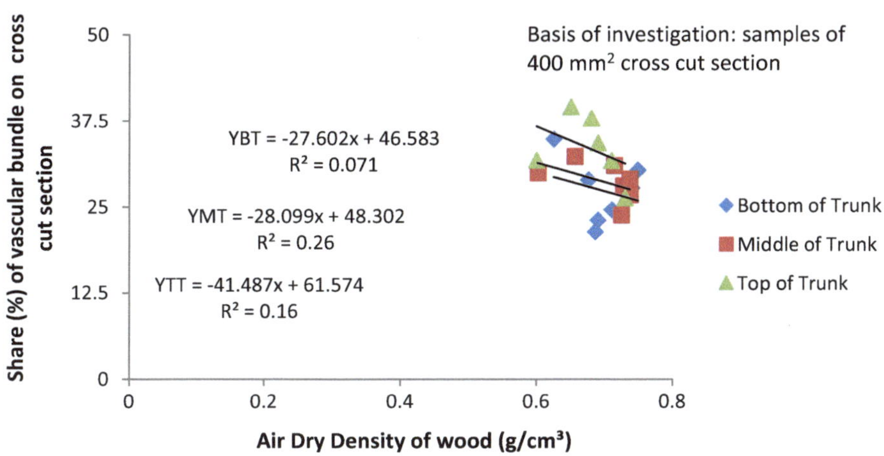

Fig. 1 Relationship between wood density and share of VBs on the cross-cut section

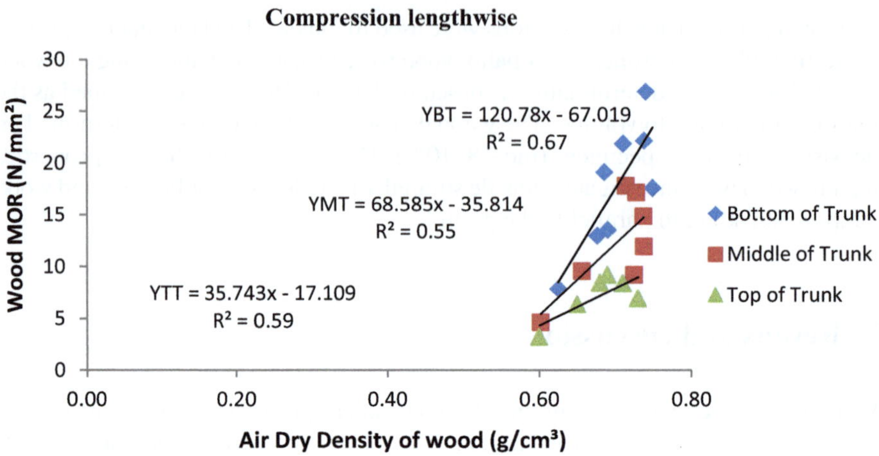

Fig. 2 Relationship between wood density and MOR

Figure 1 shows a weak correlation between the wood density and the share of VBs on the cross-cut section. The MOR in compression of wood depends positively on the wood density (medium correlation) but very weakly on the total area (percentage) of VBs (Fig. 2). Therefore, knowing only the distribution of the total area (percentage of VB from the total cross-cut area) of VBs is not enough to model the density and mechanical properties of date palm wood. It can be interpreted that, for grading lumber and for developing sawing patterns in sawmill processing, other parameters are also effective; mainly the parenchyma tissue density, chemical composition of wood, i.e., lignin, cellulose, hemicellulose, and the other anatomical parameters, i.e., proportion of fibers in a vascular bundle, cell wall thickness, and microfibrillar angle in the cell wall layers. Tension parallel-to-grain strength and tension perpendicular-to-grain are very sensitive to changes in density. The influence on tension and compression parallel to grain is influenced when loading is not exactly parallel to VB-direction (which is rarely the case as VB run not parallel to the stem axis). Also bending behavior is very much influenced by the low shear properties and tension-perpendicular to grain properties as easily shear forces and forces across the bending axis can cause failures [7].

3.1 Vascular Bundles Properties

To investigate the variation in the mechanical properties of VBs in the radial direction of date palm trunk, more than 12 VBs in subsections (1, 3, 4, 6 from top of tree and 1, 5, 9 from bottom of trunk) were separated and tested. The average strength and average Young's modulus of VBs from each zone were measured. Table 1 lists the

Table 1 Date palm: mechanical properties of vascular bundles[a]

Stem height (m)	Average tensile strength (MPa)			Standard deviation			Average elastic modulus (MPa)			Standard deviation		
	P	C	I	P	C	I	P	C	I	P	C	I
7	81	88	79	22.9	16.5	32.8	2218	3577	3540	340	400	473
2	215	269	270	48.3	65.9	67.4	14,273	17,699	18,466	2575	3553	4174

[a]Values shown for zones are based on 12 VBs; P near "the bark" of the trunk (peripheral), C between outer and inner zones (central), and I near the core of the trunk (inner)

test results for tension properties of single VBs (the MOE and the MOR). Figure 3 shows the relationship between the MOE and the MOR, respectively, and the relative radius from center.

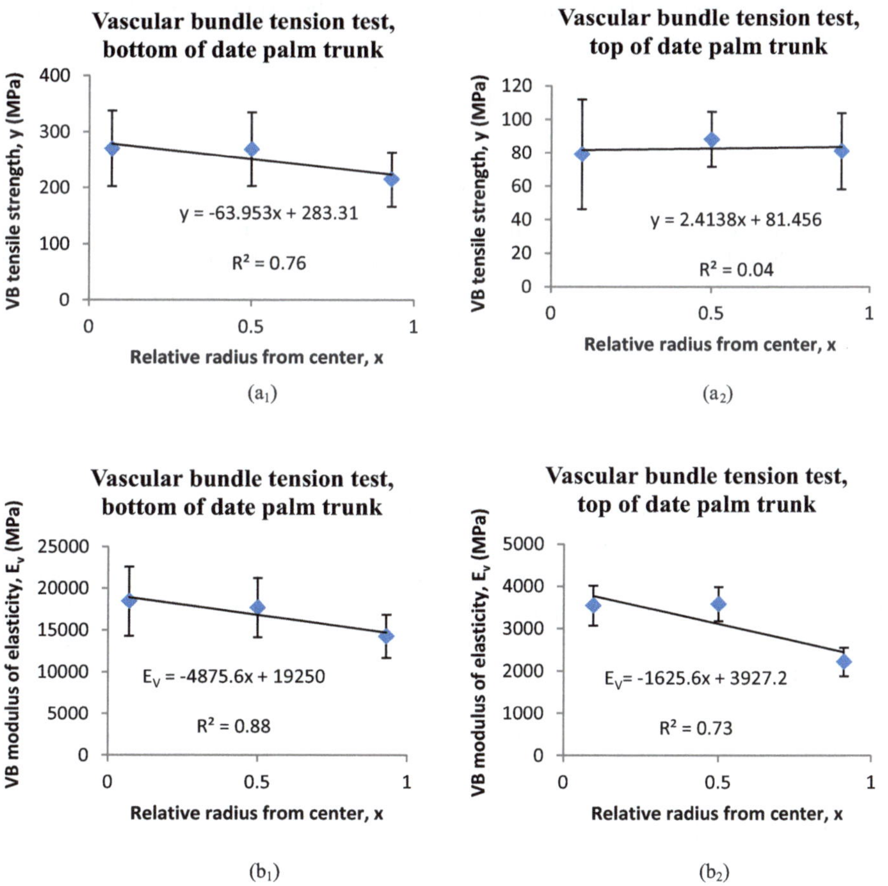

Fig. 3 Date palm: Variation in the mechanical properties of vascular bundles along the radial direction: (a_1 and a_2) tensile strength and (b_1 and b_2) modulus of elasticity

From Table 1, it becomes clear that single VBs of date palm wood in the inner zones have higher tension properties (MOE/MOR) compared to those in the outer zones. No explanation could be found for these differences, which are contrary to coconut and oil palm. No indication is given in the literature. This observation should be clarified in further research. Contrary to the variation in the properties in the radial direction of the trunk, the tension properties of single VBs at 7 m height are only 20–25% of the 2 m height (like oil palms). This can be related as for oil palm to the presence of the VBs which are composed of young cells on the top of the tree [7]. As for oil palm, the effect is much bigger along the height compared to the radius. However, the density and the strength of the wood are not different so much.

The relationship between the VB properties and the location follows a linear model. At the bottom of the trunk, the VBs near "the bark" of the trunk show weaker properties than those near the trunk core (Fig. $3a_1$) but at the top of the trunk, there is no significant difference between the inner and the outer zone of the trunk (Fig. $3a_1$). The strength variations for all three subsections can be expressed by the following formula:

$$\text{bottom of date palm trunk: } y = Ax + B$$
$$\text{top of date palm trunk: } y = A_1x + B_1$$

with y as the tensile strength of VB, and x as the relative radius from center.

The VBs near the trunk core are stiffer than those near the bark of the trunk (Fig. $3b_1$, b_2). The correlation between the MOE and the relative radius from center is linear similarly as with the tensile strength. It can be described by the following formula:

$$\text{bottom of date palm trunk: } E_v = ax + b$$
$$\text{top of date palm trunk: } E_v = a_1x + b_1$$

where E_v is Young's modulus of VB.

Table 1 summarizes the tensile strength and the MOE for the VB of two tree heights and three zones. When the MOR × share of the VB is calculated and compared against the tension strength of date palm wood [8], a close correlation between the VB-strength and the wood strength at the top of the trunk becomes obvious (Table 2 and Fig. 4). Thus, the tension strength of date palm wood on the top of the tree is dominated by the strength and the share (volume fraction, number, and diameter) of the VB. However, the statistical evaluation revealed a weak correlation ($R^2 = 0.001$) between the VB-strength and the wood strength at the bottom of the tree. This observation suggests that other factors are important and influence the tensile strength properties (MOR-t) of the wood, especially the anatomical parameters such as (i) proportion of fibers in a vascular bundle, (ii) cell wall thickness, and (iii) microfibril angle in the cell wall layers. The fiber wall structure appears to be the single most important factor that determines the wood mechanical properties under the tensile and bending stress among Calamus species [9].

Table 2 Date palm: MOR- analysis (comparison (VB MOR × share of VB) and wood MOR)

Trunk height (m)	Zones	MOR VB (N/mm^2)	Share of VB	MOR VB × share of VB	MOR wood (N/mm^2)[a]
7	P	81	0.33	27	n.a
	C	88	0.33	29	n.a
	I	79	0.27	21	n.a
2	P	215	0.33	71	66
	C	269	0.29	78	69
	I	270	0.24	65	60

[a]According to Shamsi and Mazloumzadeh [9]

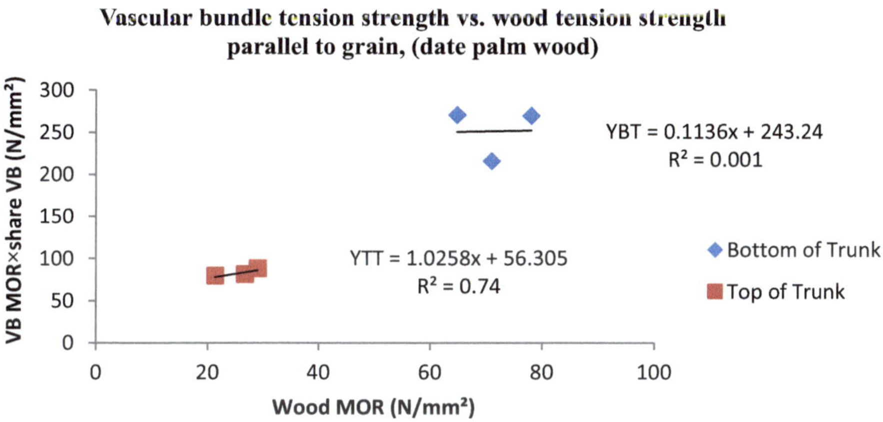

Fig. 4 Relationship between the tensile strength of vascular bundles (MOR VB × share VB) and the tension strength of wood

4 Conclusion

Based on the results, further research and development, followed by promotion, date palm wood will become a valuable commodity. Therefore, increased utilization of date palm wood may assist to conserve the other forests and supplement other wood species, which currently make up the bulk of local timber markets. Further studies are required on the mechanical properties of date palm wood, namely shear strength, bending strength, and tension perpendicular and parallel to grain strength are performed, in order to characterize date palm wood for product utilization.

References

1. Eiserhardt WL, Svenning JC, Kissling WD, Balslev H, Ann H (2011) Bot 108:1391–1416
2. Killmann W (1993) University of Hamburg, Germany, p 213
3. Fathi L (2014) Ph.D. thesis, University of Hamburg, Germany, p 248
4. Bahmani M, Schmidt O, Fathi L, Frühwald A (2014) Wood Mater Sci Eng 11:239–247
5. Jones LD (1996) Palms: throughout the world. Sci Teach 63:98
6. DIN-Taschenbuch (1991) Berlin, DIN Deutsches Institut für Normung e.V. 31, pp 87–88
7. Fathi L, Fruehwald A, Koch G (2014) Holzforschung 68:915–925
8. Darwis A, Nurrochmat DR, Massijaya MY, Nugroho N, Alamsyah EM, Bahtiar ET (2013) R. Safe'i. Asian J. Plant Sci. 12:208–213
9. Shamsi M, Mazloumzadeh SM (2009) J Agric Sci Technol 5:17–31

Mechanical Dewatering of Wet Oil Palm Lumber Prior to Press-Drying

Katja Fruehwald-Koenig, Nathan Koelli, and Arno Fruehwald

Abstract Due to shorter rotation cycles compared to other palm species, oil palm wood has lower densities, a wider density range (150–600 kg/m^3 dry), and a high moisture content. The moisture content varies, reversely to the density, with the highest values of up to 600% at the trunk core, especially at the top of the palm trunk. Kiln drying material with such low density, very high moisture content and high sugar and starch content is difficult and often produces drying defects such as cell collapse, cracks, and mold. Mechanical dewatering of wet oil palm lumber in an unheated double roller press reduces the water content and generates sugar-containing pressed water (sap) that could be used as a source for biochemicals. The share of mechanically removed water varied from 1 to >54% of the total water removed from wet to dry state (at 20 °C, 65% rh). Center boards from the top of the trunk had the highest dewatering rates while boards from the periphery at the bottom of the trunk (high density, low water content) showed the lowest.

Keywords Oil palm lumber · Mechanical dewatering · Press-drying

1 Introduction

1.1 Background

Oil palm plantations have developed rapidly because of the high demand for vegetable oil and the high productivity of oil palms. Today some 25 million hectares are under cultivation with increasing tendency. After 25–30 years of age, decreasing

K. Fruehwald-Koenig (✉) · N. Koelli
Department of Production Engineering and Wood Technology, Ostwestfalen-Lippe University of Applied Sciences and Arts, Lemgo, Germany
e-mail: katja.fruehwald@th-owl.de

N. Koelli · A. Fruehwald
Institute of Wood Science, University of Hamburg, Hamburg, Germany

M. Jawaid et al. (eds.), *Proceedings of 2nd World Conference on Byproducts of Palms and Their Applications*, Springer Proceedings in Materials 19,
https://doi.org/10.1007/978-981-19-6195-3_2

oil productivity and increasingly difficult harvests cause the plantations to be cleared and replanted. A plantation's fiber biomass at the time of clearing can amount to nearly 70 dry tons per hectare (equivalent to some 200 m^3) of trunk material and 40 tons of palm fronds (recalculated from [1]). The amount of trunk volume from clearings worldwide is between 120 and 200 million m^3 per year for the period 2010 to 2030, even more beyond 2030. For almost 50 years, palm trunks and their fibers have attracted scientific and commercial interest [2–5]. The volume of oil palm trunks (OPT) makes it an appealing supplement to, or even substitute for, tropical wood that is increasingly in short supply. Research has been conducted on the use of OPT as a replacement for other types of timber, energy, and chemicals [2, 3].

Early research concentrated on material properties (structure, density, mechanical and chemical properties), later also focused on processing pilot products. The reason why oil palm wood (OPW) is rarely used and has no market relevance until today originates in the material itself and the lack of adapted manufacturing processes and innovative product development. OPW is structured by a matrix of thin walled, soft parenchyma cells with low overall density (0.1 to 0.25 g/cm^3), hardness and strength, reinforced by vascular bundles with high density (0.7–1.2 g/cm^3 [6]). This extreme inhomogeneity in the anatomical structure, the density and property gradient within the trunk and the anisotropic behavior makes the material extremely difficult to process with common wood processing techniques in regard to tool wear, rough surfaces, splitting, and grading [7–10]. High moisture and high sugar and starch content in the trunk, especially in the parenchyma matrix, make it extremely difficult to dry the wood. If traditional drying technique and kiln operation is applied, drying defects like cell collapse (Fig. 1), splits, cracks, mold, discoloration, and sticker imprints can easily occur. The risk of cell collapse when drying oil palm lumber is especially high as reported for example by [4, 5, 11, 12].

In general, cracks, splits, cell collapse, and sticker marks/imprints are common problems when drying low-density wood with high moisture content. Influencing parameters are (1) drying temperature as the wood substance becomes softer (due to lower mechanical properties of the wood material and more plastic deformations), (2) high drying ratio due to low relative humidity in the kiln and high air speed along the wood surface which increases evaporation of water from the wood surface.

It is well known how cell collapse develops, i.e., [13–17]: Wood tissue consisting of cells with lumen filled with water is dewatered by a flow of liquid water driven by

Fig. 1 Typical cell collapse in oil palm wood after improper kiln drying

evaporation of water at the wood surface. The higher the evaporation rate, the higher the flow of liquid water. The water flow creates a negative pressure/vacuum in the cell lumen that becomes more pronounced as the lumen is filled with more water. Lower water content means also the presence of air in the lumen; the air expands and the vacuum is reduced. The vacuum increases as long as the lumen is intact in its size. At a certain vacuum, the cell walls cannot withstand the vacuum and collapse. The collapse either results in a crack along the middle lamellae of two cells that is the starting point for a crack, or if the middle lamellae are strong enough, the neighbor cells are displaced from their original location. This is visible by abnormal dimension changes (Fig. 1). The thinner the cell walls are the more severely pronounced the cell collapse is. OPW from the interior can have dry density as low as 0.1–0.15 g/cm^3. Considering the vascular bundles with densities of 0.8–0.9 g/cm^3 (dry) [6, 18], the density of the parenchyma tissue can be as low as 0.05 g/cm^3 (dry) [6]. Cell walls in this case are very thin (5–8 μm at the bottom periphery and 2–8 μm at the central part [19]) and very soft in stability considering the high moisture content of the cell walls. As far as the wood itself is concerned, the cell wall thickness and strength is important. 60 vol% of the OPT show a density between 0.2 and 0.4 g/cm^3 and 20 vol% even shows a density below 0.2 g/cm^3 [21] dry with extremely thin parenchyma cell walls with low strength. Cell collapse mainly occurs during the first drying phase. The movement of the liquid water is limited by the water permeability of the wood and therefore a vacuum can form in the cell lumen. A very low drying speed with drying ratios only slightly above 1.0, low temperatures and low air speed are recommended to avoid cell collapse. This results in prolonged drying times and higher costs, but a risk for cell collapse damage still remains. A common problem in the industrial drying is that the kilns are not designed to run at extremely high relative humidity such as >90%. Leakages and vapor condensation at the kiln walls limit the necessary drying conditions, especially the relative air humidity. For some wood species that are prone to mold development, like OPW, a minimum drying temperature of above 40–45 °C is necessary to avoid mold in the kiln. Even then the high humidity in the kiln is a must to prevent cell collapse. For wood species prone to cell collapse often pre-air-drying at low temperatures is recommended [14], but also with the lower temperatures during air drying, a high air humidity is needed to avoid cell collapse. For oil palm lumber (OPL) [12] found mold problems during air drying in tropical climate and still some cell collapse. For OPW in lumber dimensions, a balance between temperature, humidity, and ventilation/air speed is necessary to minimize drying defects like collapse, cracks and mold or discoloration development. For OPW in veneer dimensions cell collapse is normally not a problem since the low thickness of the veneer allows good water movement even at higher drying rates [21].

1.2 Research Approach

One strategy to overcome the drying problems of OPW in terms of drying time, costs and defects could be the development and use of new drying strategies like adopted drying schedules, combined air-/kiln drying, separate kiln load according to wood density (moisture content), kiln design and kiln operation. Another technical solution to avoid cell collapse could be to replace a part of the free water in the cell lumina with air by mechanical pre-dewatering followed by adopted kiln drying [22]. Here the air bubble inside the cell lumen expands while drying, the built-up of a vacuum is lower and cell collapse can be reduced or even avoided. The liquid water in the cell lumina can be reduced by applying a certain load on the OPW, which generates an overall pressure inside the wood and the water flows to the wood surfaces according to the anisotropic water permeability [23]. After reducing the load, the wood moves back towards its original size (spring back) and air is sucked into the wood.

In research, hydraulic presses or testing machines are used in most cases to investigate the mechanical dewatering of small wood samples by flat compression [22, 24, 25]. In practice, the cold mechanical dewatering by flat compression of lumber with large dimensions is limited because of long drying times caused by low permeability [26]. The thermo-mechanical press-drying has greatly reduced drying times compared to kiln drying, but presents the risk of honeycomb formations [15]. The press drying process can be combined with the densification of the wood.

The press plate is flat and smooth because a profiled press plate surface causes deep imprints in the wood surface and wood particles are blocking the grooves in the plate. In a daylight press, the mechanical dewatering of lumber up to 4 m length along the fiber direction is low. Most of the water dewaters over the board edges. Therefore, dewatering in a daylight press takes quite long and high pressure is needed. In contrast, in a roller press the lumber is dewatered between two rollers of a certain diameter. The pressure is applied only on a small surface (line) depending on the roller diameter and roller penetration depth. The water flow is mainly in longitudinal direction of the board in front of the roller pair in the form of "a water wave". The water pressure forces the water towards the uncompressed end of the board in front of the rollers, where the water is pressed out at the board faces. It could be helpful to remove the water in front of the rollers to avoid soaking it back into the wood, when the "water wave" moves into the rollers gap.

1.3 Research Goal

The basic idea and core strategy to improve the drying process of oil palm lumber (OPL) is mechanical (cold) dewatering followed by either kiln drying or press drying in a hot daylight press. The squeezed out water is replaced by air when the wood moves back towards its original size (spring back) after pressing and the air in the cells reduces the risk of cell collapse and severe cracks in the subsequent kiln drying. Often, especially for low density material, the board surfaces are quite rough after mechanical dewatering in a roller press and subsequent spring back and kiln drying, which requires intensive surface processing (planing/sanding) of the dried lumber with high costs and low yield. Press drying in a heated daylight press takes less time after mechanical dewatering of the boards and shows very smooth board surfaces.

The squeezed out sap from mechanical dewatering before drying contains sugars and starch which can reduce the mold and fungi infection during drying and later use of the kiln-dried wood and the sugars and starch may be used for producing biochemicals as well.

2 Material and Methods

2.1 Material Procurement and Sample Preparation

20 oil palms (*Elaeis guineensis* J_{ACQ}. var. *Tenera*) were push-felled 0.5 m above ground in a 30-year-old plantation near Labis in the state of Johor, Malaysia, in March 2020. The trunks were cut by chain saw into two sections of 5.2 m each (Fig. 2). The trunk sections were transported to a nearby plant, treated with gas against insects and 16 bottom and 5 top sections were loaded in a reefer container (–10 °C, time between felling and freezing approx. 3 days) and shipped to Germany.

In a sawmill, the 5.2 m trunk sections are cut into two trunk parts of approx. 2.0 m length and into trunk disks for further investigation. The trunk parts are processed into boards on a band saw (in frozen state) according to the sawing pattern in Fig. 3. The boards are distinguished in "outer" (no. 1, 3, 5, and 7), intermediate—"inter" (no. 2, 4, 6, and 8) and "inner" boards (no. 9, 10, 11, and 12). Outer and intermediate boards are cut to 33 mm thickness, while inner boards are cut to 50 mm thickness. Outer and intermediate boards are cut thinner than the inner boards, because oil palm trunks show a high-density gradient from the periphery in the radial direction following power function with densities of approx. 200–600 kg/m^3 dry material [20, 27]. Therefore, the density gradient is more pronounced in the outer and intermediate boards, especially in thicker and wider the boards. Bakar et al. [28] developed an advanced polygon sawing pattern for oil palm trunks which is oriented on the density gradient. However, for practical and scientific (groups of identical boards) reasons, the conventional sawing pattern for band saws is used (Fig. 3). This sawing pattern is easy to implement and produces two identical (mirrored) boards for each type of board (outer, inter, and inner). This is important because 50% of the boards (with numbers 3, 4, 7, 8, 9, 10) are kiln dried immediately after sawing. These boards serve

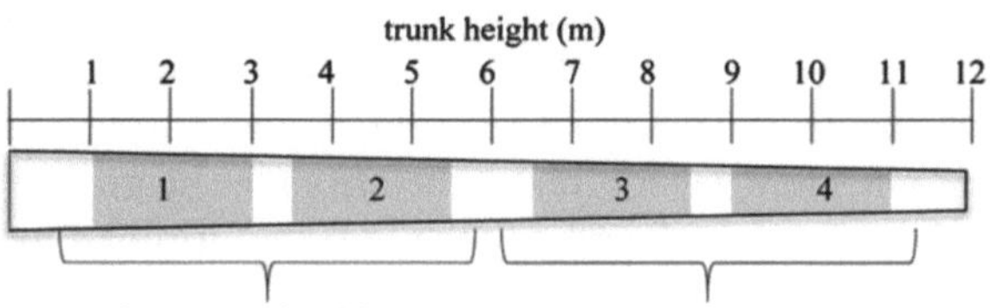

Fig. 2 Breakdown of the oil palm trunks with 4 trunk parts

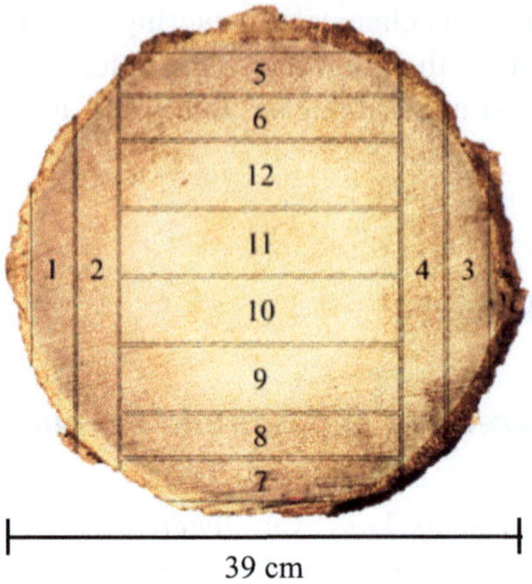

Fig. 3 Applied sawing pattern with board numbering according to the order of sawing

as reference samples for the comparison of the mechanical properties of densified and non-densified wood (not included in this publication). The other 50% of the boards (with numbers 1, 2, 5, 6, 11, 12) are prepared for densification and wrapped in plastic foil, transported to the laboratory of Ostwestfalen-Lippe University and stored in a refrigeration chamber at –10 °C. In total 47 boards were dewatered with a roller press. From the bottom section of four palms 13 boards were thermo-mechanical dewatered and 22 boards were cold dewatered. From the top section of one palm six boards were thermo-mechanical dewatered and six boards were cold dewatered.

2.2 *Mechanical Dewatering with a Roller Press*

The boards are taken out of the refrigeration chamber 24 h before cold compression and stored wrapped at 20 °C. The compression is realized on an unheated glue roller machine (built by company Bürkle, Freudenstadt, Germany; model DAK 1300) that was modified to a roller press with one pair of rollers (Fig. 4). The steel rollers are rubberized with slight circumferential grooves (up to 1 mm depth) and have a diameter of 240 mm. The rollers are the only feeding device through the machine with feeding speeds of 6–30 m/min and set to 12 m/min. At a feeding speed below 12 m/min overload leads to a standstill of the rollers, because the indentation depth of the rollers in the boards is deeper and hence more power is required. At a feeding speed above 12 m/min, more water is reabsorbed from the board face since the water has less time to drop from the board surface into the drip pan that is positioned underneath the rollers.

Wingate-Hill and Cunningham [24] examined the mechanical dewatering of small blocks of sapwood from four commercial timber species in a daylight press and report slightly higher water losses for slower compression speed (0.4…8…15 mm/min) and

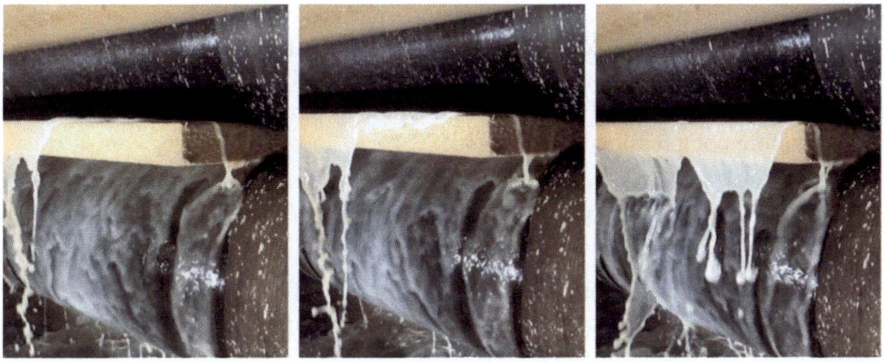

Fig. 4 Compression of an intermediate board with roller press

for longer load time. The negative correlation between moisture content reduction and compression speed was also reported from Zhao et al. [25] for Chinese fir and poplar for compression speed between 0.5 and 10 mm/min for small specimen of 3, 5 and 10 cm. In a roller press the compression speed depends much more on the roller diameter, the degree of compression, and the feeding speed. Adachi et al. [29] stated, that the amount of mechanically removed water of a veneer in a roller press depends mainly on the compressive deformation (15–75%) and does not depend on the feeding speed (1–10 m/min).

Within this research, the applied pressure on the boards is defined by the weight of the top roller unit of 7650 N. The line pressure varies because the width and the indentation depth varies between the boards depending on the boards' resistance against deformation, which mainly depends on its density. The compression is not distance-dependent, because the top roller unit is not fixed at the adjusted gap, but is elevated due to the pressure resistance of the board when the board passes through the rollers. The roller distance is adjusted in three steps for three subsequent throughputs (25, 20, and 15 mm for boards with 33 mm initial thickness and 40, 30, and 20 mm for boards with 50 mm initial thickness), even though the preset distance is not reached due to the flexible top roller. The degree of compression (DC) is the percentual thickness reduction from the initial thickness (IT) to the distance (D) in the press nip based on the initial thickness (IT) (Eq. 1).

$$DC = \frac{\text{IT-D}}{\text{IT}} \cdot 100[\%] \tag{1}$$

Since the set distance is not reached but the top roller is elevated, the DC is estimated to 20–50% with higher compression of the inner (softer) boards. The boards' thickness after compression cannot be measured exactly because of the soft flexible state, the rough surface, the unequal deformation, and the spontaneous spring back. The mass of the board is weighed before and after each passage through the roller press. Compressed air is used to remove the water from the rollers on the inlet side and from the board surface to prevent the squeezed out water from being soaked into the board again. The dewatering rate (DR) in the roller press is calculated as the percentual reduction of the initial moisture content (IMC) before compression to the final moisture content (FMC) after three passages based on the initial moisture content before compression (Eq. 2).

$$DR = \frac{\text{IMC-FMC}}{\text{IMC}} \cdot 100[\%] \tag{2}$$

The IMC is calculated with the wet mass ($m_{u\text{-initial}}$) of the boards before the roller press and the calculated dry mass (m_0). For the calculation of the FMC, the wet mass ($m_{u\text{-final}}$) after the roller press is used (Eq. 3).

$$MC = \frac{m_u - m_0}{m_0} \cdot 100[\%] \tag{3}$$

The dry mass (m_0) of the boards is calculated after drying in the heated single-daylight press (cf. Sect. 2.4) and conditioning the boards to 20 °C and 65% rh. After conditioning, samples are taken from the boards and oven-dried at 103 ± 2 °C according to [30] until constant weight to determine the moisture content at the conditioned state ($MC_{20/65} = 9\text{–}13\%$). Drying tests showed that oven-dried (103 ± 2 °C) oil palm samples stored in the dry oven at temperatures of 150 °C for 2–4 h have a mass loss of approx. 4%. This is assumed to be the maximum mass loss of the boards during drying, and therefore, m_0 was calculated by the dry mass multiplied by 1.04 (Eq. 4).

$$m_0 = \frac{m_{u20/65}}{(MC_{20/65} + 1)} \cdot 1.04[g] \tag{4}$$

The extracted water (EW) is calculated as difference of the mass before ($m_{\text{before RP}}$) and after compression on the roller ($m_{\text{after RP}}$) press divided by the volume before compression on the roller press ($V_{\text{before RP}}$) (Eq. 5).

$$EW = \frac{m_{\text{before RP}} - m_{\text{after RP}}}{V_{\text{before RP}}} \left[\text{kg/m}^3\right] \tag{5}$$

2.3 *Thermo-Mechanical Dewatering with a Roller Press*

The 19 thermo-mechanical dewatered boards are steamed before the mechanical dewatering in a steam chamber (with a steam temperature of 100 °C) until the board core temperature was at least 90 °C for 1 h. After removal from the chamber, the boards are weighted and immediately compressed in the unheated roller press as described in Sect. 2.2. Steaming did not increase the moisture content of the boards and therefore did not lead to higher wet mass of the boards.

2.4 Thermo-Mechanical Press-Drying in a Single-Daylight Press

After the cold or thermo-mechanical dewatering on the roller press, all boards are thermo-mechanical press-dried in a single-daylight press (built by company OTT, Lambach, Germany; model Stabil) for 4–8 h at temperatures of 120–150 °C with a pressure of 0.5–1.2 MPa. The single-daylight press has a pressure-dependent control, which is adjusted stepwise until the aimed distance with spacers is reached. Before the compression, the boards are heated up in the closed press until the board core temperature of 80 °C is reached. The water removal in the single-daylight press occurs in two overlapping stages, firstly liquid water is removed mechanically by the compressive force and simultaneously with increasing temperature, more water is removed thermally by evaporation. The board core temperature stays around 100 °C until the remaining water is evaporated. Then the board core temperature rises to the set temperature and therewith indicates that the moisture content of the boards is very low. The boards are removed from the press and put in a cooling stack between OSB panels for 24 h. Subsequently, the boards are conditioned in the climate chamber at 20 °C and 65% rh.

3 Results and Discussion

3.1 Reduction of the Moisture Content Achieved by Mechanical Dewatering in the Roller Press

A comparison of the cold and hot dewatered boards is given for each location within the trunk (Fig. 5). The full-color columns show the reduced moisture content after mechanical dewatering, and the shaded area shows the decrease of moisture content by cold or thermo-mechanical dewatering in the roller press. Stacked together the full-color column and shaded area show the initial moisture content of the board. Black columns represent the thermo-mechanical and grey columns the mechanical (cold) dewatering. The numbers at the bottom of the columns indicate the number of the boards from which the mean value was calculated, the error indicators show the S.D. The moisture content is dependent on the position within the trunk. With increasing trunk height and distance from the cortex, the amount of free water and consequently the moisture content increases. The mean moisture content of the boards varies between 160 and 270% in the outer boards, between 210 and 400% in the

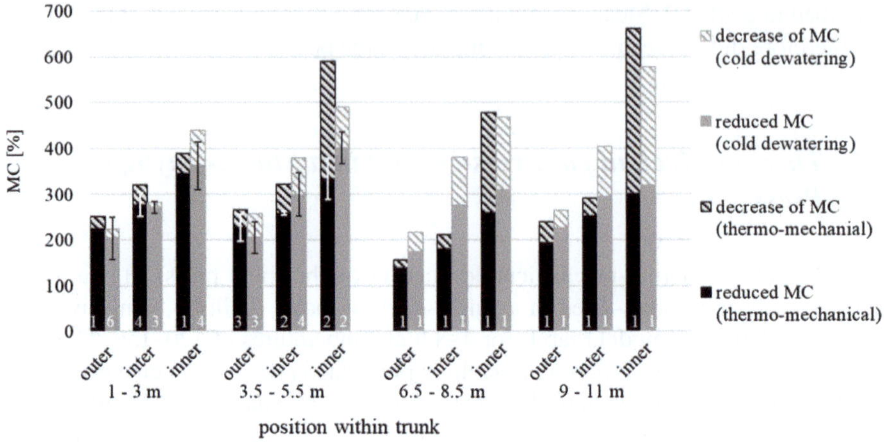

Fig. 5 Reduction of moisture content of thermo-mechanical and cold dewatered boards

intermediate boards, and between 390 and 660% in the inner boards. The wide-ranging moisture content values are slightly higher but comparable to those reported in [20, 27].

The amount of pressed water decreases with each passage through the machine (Fig. 6). In the first passage of the thermo-mechanical dewatered boards 47.6% (S.D. 10.0%), in the second passage 26.9% (S.D. 7.8%), and in the third passage 25.6% (S.D. 11.2%) of the squeezed water is removed from the boards. In the first passage of the cold dewatered boards 53.9% (S.D. 12.2%), in the second passage 27.0% (S.D. 8.2%), and in the third passage 19.1% (S.D. 6.3%) of the squeezed water is removed from the boards.

The inner softer wood contains more free water (higher IMC) and has lower wood density and therefore, is easier to squeeze. The dewatering rate depends on the IMC especially for the thermo-mechanical dewatered boards (Fig. 7). The highest

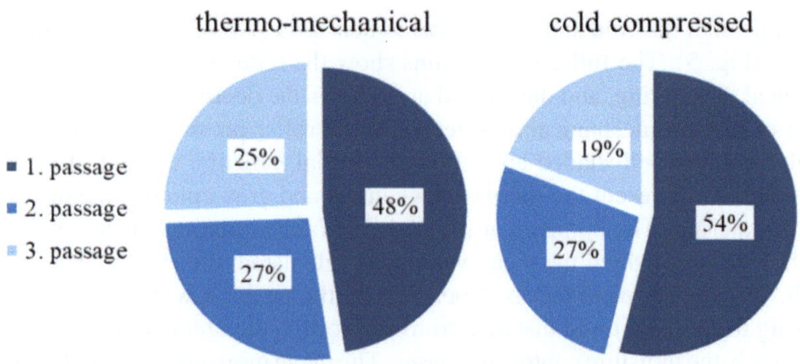

Fig. 6 Share of pressed water on the passages through the roller press

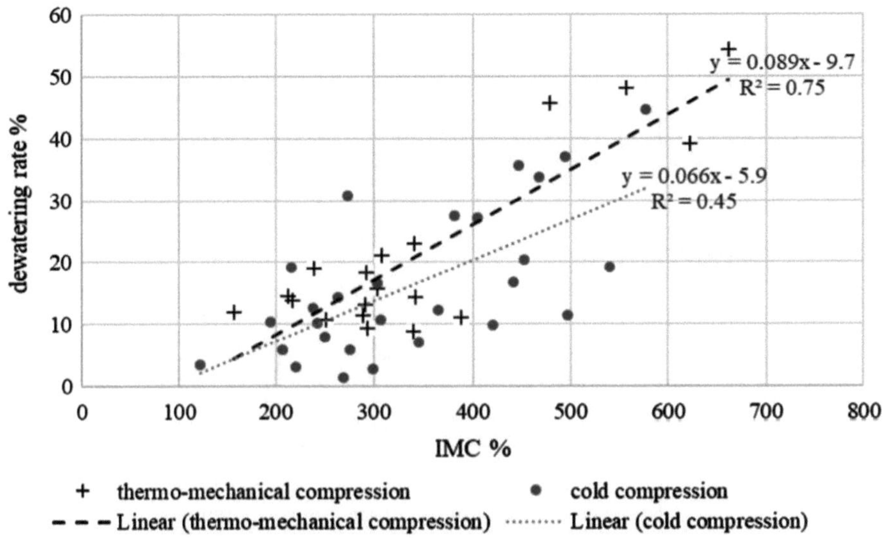

Fig. 7 Relationship between dewatering rate and IMC

dewatering rates are in the inner boards from the highest trunk part (Fig. 5). The dewatering rate varies between 1 and 45% for cold dewatered and 9–54% for thermo-mechanical dewatered boards.

The multiple regression analysis for the DR and the two independent variables, x_1: position within cross section (distance to cortex in mm) and x_2: position within height (distance to ground in m) results in Eq. (6) with an adjusted coefficient of determination of $R^2 = 0.51$. Statistically, there is no significant difference between cold and thermo-mechanical dewatering, and therefore, only the position within the cross section and position within height is used to calculate the equations.

$$DR = (0.13 \cdot x_1) + (2.36 \cdot x_2) - 3.5 \tag{6}$$

The multiple regression analysis for the amount of extracted water EW (kg/m^3) and the two independent variables x_1 and x_2 results in Eq. (7) with an adjusted coefficient of determination of $R^2 = 0.51$.

$$EW = (1.1 \cdot x_1) + (19.49 \cdot x_2) - 38.7 \tag{7}$$

Figure 8 shows a schematic of how the dewatering rates develop (left side) when calculated with Eq. (5) and how much extracted water (right side) appears (kg/m^3) when calculated with Eq. (6) for the different positions of the middle of the board within the trunk height and distance to cortex.

height [m]	dewatering rate [%]					extracted water [kg/m³]				
12	30	36	41	46	52	417	373	329	284	240
10	26	31	36	42	47	378	334	290	245	201
8	21	26	32	37	42	339	295	251	206	162
6	16	21	27	32	38	301	256	212	167	123
4	11	17	22	27	33	262	217	173	128	84
2	7	12	17	23	28	223	178	134	89	45
	40	80	120	160	200	200	160	120	80	40
	distance to cortex [mm]					distance to cortex [mm]				

Fig. 8 Dewatering rates in % and extracted water in kg water per m^3 wet board for different positions within the trunk calculated with Eqs. (5) and (6)

3.2 Influence of Mechanical Dewatering on the Surface and Shape of the Boards

The rubberized rollers with circumferential grooves leave flat markers on the surface of the inner boards (Fig. 9b). The outer and intermediate boards often show cracks lengthwise close to the longitudinal edge (Fig. 9c) while some of the inner boards tend to twist (Fig. 9d). With reducing press nip and repeated passages through the roller press, the board edges become rough and loose and more and more vascular bundles stick out of the surface (Fig. 9d). Steaming before compression shows a positive effect on the board's quality because cracks are less pronounced in the thermo-mechanical dewatered boards.

In pre-tests boards are compressed with an unheated single-daylight press and afterwards kiln dried with pressure on the stickers (Fig. 10). Where pressure is applied, the surface becomes smooth, the thickness lower and the density higher and the board is less warped and twisted. In between the stickers, where no pressure is applied, the shape remains loose and many vascular bundles stick out the surface. This is like the appearance of (unwanted) sticker marks during conventional drying, that can occur when oil palm boards are kiln dried in stacks and too much load is applied per square unit either when the number of stickers or their size is too low (too small total sticker area).

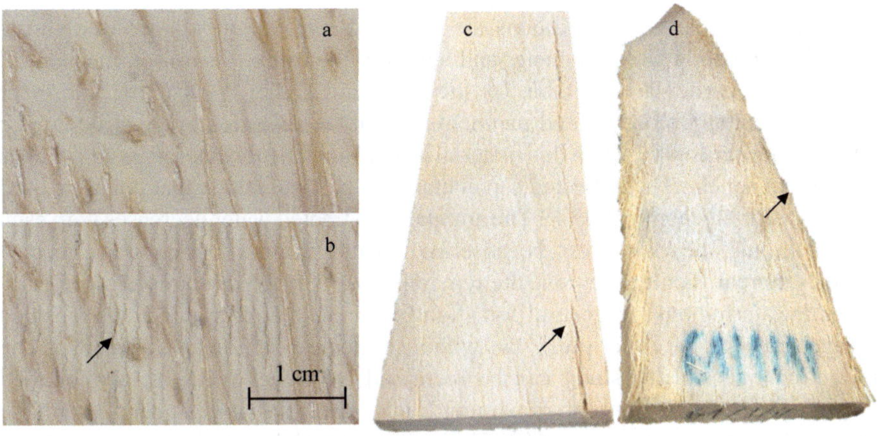

Fig. 9 **a**, **b**: Surface quality before (**a**) and after (**b**) the roller press of an inner board with small flat markers (arrow); **c**: Intermediate board with lengthwise crack (arrow); **d**: twisted shape of an inner board with loose edges (arrow)

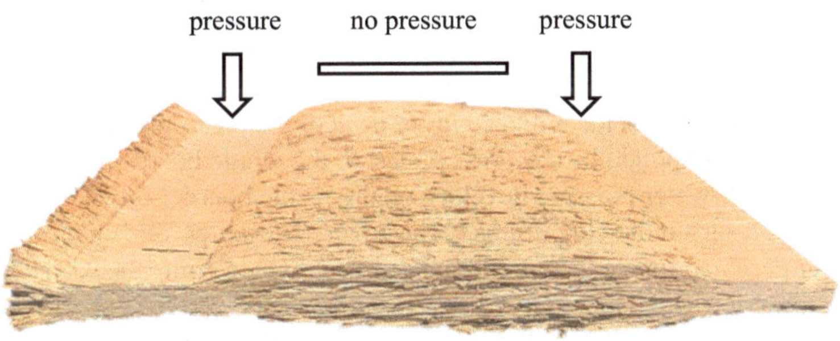

Fig. 10 Partially compressed board during kiln drying

4 Discussion and Conclusions

Mechanical dewatering reduces the high initial moisture content of oil palm boards by removing a certain amount of the free water in the roller press. The amount of extracted water/dewatering rates for different boards shows a strong dependency on the board location within the trunk. Lowest dewatering rates are found in the outer part of the bottom section, while the inner part has the highest dewatering rates, in particular at the top of the palm. Since palms do not possess a cambium ring, they only grow with their apical meristem upwards and hence have younger and softer cells at the top [31]. Density and compressive behavior (elasticity, plasticity, strength) decrease from the trunk periphery inwards [32] while the moisture content rises in reverse direction [20, 33]. Boards from the inner and higher trunk parts are less resistant against compression and contain more water than boards from the outer and lower trunk parts. This explains the observed correlation between dewatering rate and the position along diameter and height of the trunk. Therefore, mechanical dewatering is particularly suitable for the inner and higher trunk parts. The inner trunk parts contain more ground parenchyma that serves as a storage structure for water and nutrients [31], when this material is mechanically dewatered large volumes of press sap appear. The collected sap contains sugars and starch that could be used as a source for biochemicals [34]. The amount of extracted water decreases with each passage through the roller press, because the moisture content decreases and hence it becomes more difficult to remove the remaining sap. Steamed boards show slightly higher dewatering rates in general and slightly more extracted water in the repeated passages through the roller press, but without statistical significant difference. The dewatering rates in this study can be increased by higher pressure. Compressed wood chips produced from the remaining trunk core from veneer peeling could be dewatered mechanically up to 80% [35]. But the optimal dewatering rates depend on the dimensions and the application of the material, because very high compression (for fast drying and maximum extraction of sugar-containing press sap) may compete with the board quality.

The compression in the roller press is a continuous process in which pressure is applied continuously on a small surface ("line") at a time. Inside the press nip, the water is forced to drain to the boards edges or further lengthwise into the board in front of the rollers. The water pressure in the board on the inlet side rises and water is pressed out of the boards faces and edges in front of the press nip. In a flat daylight press the pressure is applied by the press plates on the whole board surface simultaneously and a high-water pressure develops on the whole board. Forced by the water pressure the water from the core drains to the edges. The water movement is limited by the water permeability of the wood, that decreases when compressed. Therefore, slower compression speed and higher press force is needed when compression takes place in a daylight press. In a roller press, the water drains from the compressed area to the still uncompressed area where the water permeability is higher, therefore much higher compression speed is possible. The compression speed in the roller press depends on the feeding speed, the roller diameter, and the degree of compression. The high feeding speed of 12 m/min used in this study was the slowest possible that could be realized. If applied on an industrial scale further investigations about the optimal feeding and compression speed with roller press should be investigated.

Before the water is removed through mechanical dewatering resp. compression of wood, the existing air in the wood is removed first due to the higher viscosity of water [29]. Consequently, the mechanical dewatering rate under defined conditions depends on the degree of water saturation of the specimen. OPW naturally has higher moisture content and less air than common wood species that are rarely water saturated. Therefore, [24, 25, 29] increased the moisture content of the investigated specimen to a fully saturated state before compression. According to the equation from Kollmann [37, p. 392], the maximum moisture content (in decimal number) can be calculated with $\text{m.c. max} = 0.28 + ((1.50 - \rho_0)/(1.50 \cdot \rho_0))$. Gan et al. [28] reports a density range of the basic density (m_0/V_u) in 30-year-old oil palms between 0.141 and 0.635 g/cm^3. The saturated moisture content range calculated with these figures is between 1.19

and 6.71 (119–671%). This leads to the assumption that the boards used in this study with moisture content values between 160 and 660% are almost fully water saturated (containing almost no air). If fully saturated boards are mechanically dewatered in a roller press, the volume is reduced temporarily at the press nip and returns to almost the original thickness after the roller pair (spring back). The emerging cavities are filled with air and water that is undesirably soaked from the press nip in the feed direction. To prevent the water from being soaked back into the board behind the rollers at the outlet area, the board should be held to constant thickness behind the rollers. This could be realized by a press with a continuous press belt familiar in particleboard production.

The risk of cell collapse during kiln drying is reduced when less liquid water is captured inside the wood because the negative evaporation pressure is equalized by air, which, unlike water, is able to expand. Rittiphet et al. [22] found that internal voids on kiln dried OPW disappear if the samples are compressed to 35–55% of the initial thickness before drying, but cracks are still visible at the cross-cut sections of the samples. In this study, the compressed boards were further thermo-mechanical press-dried. Pretests showed that a long-lasting thickness reduction is only reached by press-drying. Drying without load on the whole board surface leads to an uncontrolled spring back, a loose surface, and occasionally distortion of the boards. Another advantage of press-drying is that additional liquid water can be removed mechanically.

The compression in the roller press leads to defects of the material, which increases with increasing degree of compression and (in the case of this investigation) with repeated passages through the machine. The vascular bundles follow a spiral course inside the oil palm trunk [31], which might explain the twisting of some inner boards after compression. The parenchyma is compressed and inner growth tensions (caused by the vascular bundle orientation) become free. The vascular bundle system is relocated within the board during compression due to soft parenchyma and tensions within the board might occur. These tensions are set free during spring back and might distort the boards. The cracks in the outer and intermediate boards often occur in the area with a sharp density gradient, which was observed by Koelli [21] in the first few centimeters from the trunk periphery. The resistance of the OPW against compression load perpendicular to the vascular bundles mainly depends on the density [33] and therefore cracks occur due to different deformation behavior in the density transition area. Sometimes, the surface is marked by grooves from the rubberized rollers. The small grooves somewhat limit the width expansion of the board, but the edges become fibrous and loose because there is no lateral restraint. Steaming before compression softens the boards because of the plastification of wood under heat and moisture and therefore facilitates compression. Hot compressed boards also show less cracks behind the roller press than cold compressed boards.

The compression with a roller press is easy, cheap, and quick to reduce the high initial moisture content of oil palm boards and to generate high amounts of sugar-containing sap. Mechanical dewatering with a roller press before drying reduces drying time and costs, because the energy consumption of mechanical dewatering is much lower than that of thermal water removal [37]. A mechanical dewatering for oil palm boards is possible like the mechanical dewatering of oil palm veneers by a roller press with a downstream veneer dryer [38]. A machine design with several pairs of distance-controlled rollers that work as integrated feed and form a reducing press nip can lead to higher dewatering rates under moderate compression conditions.

Acknowledgements The project was funded by the German Federal Ministry of Education and Research through the "Bioökonomie International 2017" project "Oilpalmsugar (031B0767A)". The procurement of the oil palm material was supported by the Malaysian Timber Industry Board (MTIB) and the Fibre and Biocomposite Development Centre (FIDEC) represented by Dr. Loh Yueh Feng as well as Profina Plywood Sdn Bhd represented by Chuah KK and Chuah KH.

Author Contributions Prof. Dr. Arno Fruehwald first came up with the idea of dewatering and densifying oil palm wood. Prof. Dr. Arno Fruehwald and Prof. Katja Fruehwald-Koenig together developed the research questions, the project goal, and the research approach. M.Sc. Nathan Koelli performed the laboratory investigations, their analyses and drew the figures and tables. Prof. Katja Fruehwald-Koenig and Prof. Dr. Arno Fruehwald supervised his work and reviewed the draft paper written by Nathan Kölli.

References

1. Astimar AA, Anis M, Kamarudin H, Ridzuan R, Mat Soom R, Wan Hassan WH (2011) Chapter 28: Development in oil palm biomass utilization further advances in oil palm research (2000–2010). In: MPOB 2011, vol 2, pp 896–929
2. Khozirah S, Khoo KC, Ali ARM (1991) Oil palm stem utilization—review of a research. Forest Research Institute Malaysia (FRIM), 120 p
3. FRIM-UNEP (2012) UNEP. Converting waste oil palm trees into a resource, ed. Program 2012: Collaborative Project Report, U.N.E. Program, 200 p
4. MTIB (2015) Palm timber: a new source of material with commercial value, 87 p
5. Fruehwald A (2017) Presentation on drying of oil palm lumber
6. Heister L (2022) to be published 2022
7. Sukumaran SP, Kisselbach A, Scholz F, Rehm C, Graef J (2018) Investigation of machinability of kiln dried oil palm wood with regard to production of solid wood products. Holztechnologie 59(3):19–31
8. Ratnasingam J, Ma T, Manikam M, Farrokhpayam SR (2008) Evaluating the machining characteristics of oil palm lumber. Asian J Appl Sci 1(4):334–340
9. Fruehwald-Koenig K (2019) Properties and grading of oil palm lumber. In: 21st international nondestructive testing and evaluation of wood symposium; General Technical Report FPL-GTR-272
10. Wang S, Ross (eds) (2019) U.S. Department of Agriculture, Forest Service, Forest Products Laboratory. Madison, WI, Freiburg, Germany, 204–212 p
11. Fruehwald A, Fruehwald-Koenig K, Ressel J, Seng GK (2019) Drying oil palm sawn timber. Handbook oil palm utilization, ed. PalmwoodNet. Part 5. 2019, Hamburg, 48 p
12. Kurz V (2013) Drying of oil palm lumber—state of the art and potential for improvements. M.Sc thesis, University of Hamburg 2013. 104 p
13. Kollmann F, Côté WA (1968) Principles of wood science and technology solid wood. Springer, Berlin, 592 p
14. Innes TC (1996) Pre-drying of collapse prone wood free of surface and internal checking. Holz als Roh- und Werkstoff 34:195–199
15. Simpson WT (1983) Drying wood: a review. Drying Technol 2:235–264

16. Hawley LF (1931) Wood-liquid-relations. Technical Bulletin U.S. Department of Agriculture, 34 p
17. Siau JF (1984) Transport processes in wood. Springer series in wood sciences, Berlin, Heidelberg, New York, Tokyo, 245 p
18. Fathi L (2014) Structural and mechanical properties of the wood from coconut palms, oil palms and date palms. Dissertation, University of Hamburg, 181 p
19. Bakar ES, Sahri MH, H'ng P (2008) Anatomical characteristics and utilization of oil palm wood. In: Nobuchi, Sahri (eds) The formation of wood in tropical forest trees—a challenge from the perspective of functional wood anatomy. UPM Press, pp 161–178
20. Koelli N (2016) Density and moisture distribution in oil palm trunks from Peninsular Malaysia. B.Sc thesis, University of Hamburg, 52 p
21. Feng LY, Mokhtar A, Tahir PM, Kee CK, Samsi HW, Hoong YB (2014) Handbook of oil palm trunk plywood manufacturing. Malaysian Timber Industry Board, Kuala Lumpur, Malaysia, 147 p
22. Rittiphet C, Dumyang K, Matan N (2021) Effect of pre-mechanical compression on free water removal, drying collapses and associated internal voids of oil palm wood. Eur J Wood Wood Prod 79(4):925–940
23. Voecker J (2021) Entwicklung einer Prüfeinrichtung für die Messung der Wasserpermeabilität von Ölpalmenholz (*Eleais guineensis* JACQ.). B. Eng thesis, University of Applied Sciences and Arts, Ostwestfalen-Lippe, 89 p
24. Wingate-Hill R, Cunningham RB (1986) Compression drying of sapwood. Wood Fiber Sci 18(2):315–327
25. Zhao Y, Wang Z, Iida I, Huang R, Lu J, Jiang J (2015) Studies on pre-treatment by compression for wood drying I: effects of compression ratio, compression direction and compression speed on the reduction of moisture content in wood. J Wood Sci 61(2):113–119
26. Yong Choo AC, Tahir PM, Karimi A, Bakar ES, Abdan K, Azmi I (2013) Study on the longitudinal permeability of oil palm wood. Indus Eng Chem Res 52:9405–9410
27. Gan K, Choo K, Lim S (2001) Basic density and moisture content distributions in 30-year-old oil palm (Elaeis guineensis) stems. J Trop Forest Prod 7(2):184–191
28. Bakar ES, Febrianto F, Wahyudi I, Ashaari Z (2006) Polygon sawing: an optimum sawing pattern for oil palm stems. J Biol Sci 6(4):744–749
29. Adachi K, Inoue M, Kanayama K, Rowell RM, Kawai S (2004) Water removal of wet veneer by roller pressing. J Wood Sci 50(6):479–483
30. EN 13183-1:2002-07, Moisture content of a piece of sawn timber—Part 1: Determination by oven dry method; German version. European Committee for Standardization, Brussels, 4 p
31. Corley RHV, Tinker P (2015) The oil palm. Wiley Blackwell, 639 p
32. Srivaro S, Matan N, Lam F (2018) Property gradients in oil palm trunk (Elaeis guineensis). J Wood Sci 64(6):709–719
33. Lim S, Khoo K (1986) Characteristics of oil palm trunk and its utilization. Malaysian Forester 49(1–2):3–22
34. Dirkes R, Neubauer PR, Rabenhorst J (2021) Pressed sap from oil palm (Elaeis guineensis) trunks: a revolutionary growth medium for the biotechnological industry? Biofuels Bioprod Biorefin 15(3):931–944
35. Murata Y, Tanaka R, Fujimoto K, Kosugi A, Arai T, Togawa E, Takano T, Ibrahim WA, Elham P, Sulaiman O, Hashim R, Mori Y (2013) Development of sap compressing systems from oil palm trunk. Biomass Bioenerg 51:8–16
36. Kollmann F (1951) Technologie des Holzes und der Holzwerkstoffe, vol 2. Springer, 1050 p
37. Liu Z, Haygreen JG (1985) Drying rates of wood chips during compression drying. Wood Fiber Sci 17(2):214–227
38. Choo ACY, Tahir PM, Karimi A, Bakar ES, Abdan K, Ibrahim A, Halip JA (2015) Pre-drying optimization of oil palm veneers by response surface methodology. Eur J Wood Prod 73:493–498

[illegible] (2017) [illegible]
[illegible]
[illegible]
[illegible]
[illegible]
[illegible]
[illegible]
Wood Sci Technol [illegible]
[illegible] (1985) [illegible]
[illegible]
[illegible]
[illegible]

Glued Laminated Timber from Oil Palm Timber – Beam Structure, Production and Elastomechanical Properties

Lena Heister and Katja Fruehwald-Koenig

Abstract Oil palm timber used for load-bearing purposes such as glued laminated timber (GLT) needs to have clearly defined strength and stiffness values. Because the wood density (which correlates to elastomechanical properties) varies significantly within oil palm trunks, oil palm boards must be graded based on their density across the board width in order to homogenize and improve the properties of the final product. In this preliminary investigation, 20 beams of combined GLT with four different types of graded lamellas were produced and tested in a 4-point-bending test. The results show a correlation between density and bending strength. The characteristic strength values are achieved, and the elastomechanical properties of beams based on lamellas that are ripped lengthwise and edge-glued according to their density are higher compared to beams based on lamellas cut only according to their geometry. A correlation between bending strength and local MOE is determined. In summary, lengthwise ripping of oil palm boards according to their density across the board width as well as a grading according to density limit values is shown to improve the properties of combined GLT made from oil palm timber for load-bearing purposes.

Keywords GLT · Oil palm timber · Density · GLT production · Properties · Building products

1 Introduction

Oil palm trees (*Elaeis guineensis* JACQ.) are mainly cultivated in large plantations for palm oil production and used for food, chemicals, pharmaceuticals and bioenergy. The palm trees' oil productivity decreases after 20 years of age. Therefore, plantations are renewed every 25–30 years. Each year a large supply of oil palm trunks (some 200 million m^3 per year worldwide, with over 80% in Southeast Asia) is traditionally considered as waste. Recent research, however, has explored the potential

L. Heister · K. Fruehwald-Koenig (✉)
Departement of Production Engineering and Wood Technology, Ostwestfalen-Lippe University of Applied Sciences and Arts, Lemgo, Germany
e-mail: katja.fruehwald@th-owl.de

M. Jawaid et al. (eds.), *Proceedings of 2nd World Conference on Byproducts of Palms and Their Applications*, Springer Proceedings in Materials 19,
https://doi.org/10.1007/978-981-19-6195-3_3

commercial uses of oil palm wood [1]. In many cases the wood can substitute for tropical hardwoods, e.g. used as panels (blockboards, flash doors, multi-layer solid wood panels) and softwoods in construction timber (GLT, CLT). Only few studies were made on building products: [2] tested rafters with small binders and [3] tested the compression properties of oil palm wood CLT. Numerous studies show that the elastomechanical properties of oil palm wood vary with the density along the height and diameter of oil palm trunks [4–11]. Therefore, conventional sawing patterns lead to density and further strength and stiffness gradients within the cross-section of boards. In order to use oil palm timber for load-bearing products with defined elastomechanical properties, strength grading must be carried out. [12] showed that existing X-ray devices, if properly calibrated, can be used for oil palm lumber. The aim of this preliminary investigation is to determine elastomechanical properties of GLT made from strength graded oil palm wood. Therefore, 20 beams of combined GLT from oil palm wood with four different types of strength graded lamellas were produced and tested in a 4-point-bending test. These initial tests intend to show that, firstly, grading according to density limit values and the specific arrangement of the lamella within the glulam beam and, secondly, lengthwise ripping of oil palm boards according to their density across the board width both lead to an improvement of the elastomechanical properties of GLT made from oil palm lumber for load-bearing purposes.

2 Material and Methods

2.1 Oil Palm Wood Material

The material was taken from 30-year-old oil palms (*Elaeis guineensis* JACQ.) grown near Kluang/Johor, Malaysia. In 2016, a total of 150 palms were harvested for sawmilling and drying studies by PalmwoodNet and the Forest Research Institute of Malaysia (FRIM). The sawing pattern is shown in Fig. 1. The denser material in the periphery was cut to 30 mm (fresh) and had a thickness after kiln drying of 27 mm, the lower dense material from the centre was cut into 55 mm (fresh) resp. 50 mm (kiln-dried) boards. The kiln-dried material was shipped to Germany. For the production of GLT, 70 boards of 27 mm thickness from 20 different logs and two different trunk heights were taken. The lower trunk section was from 1–4 m and the middle trunk section from 4–7 m trunk height above ground. The mean length of the boards was 2.8 m and width 0.22 m.

Due to the anatomical structure of monocotyledons, it was assumed that opposite boards have comparable elastomechanical properties. For the pairwise comparison in this investigation, two outer and two inner 27 mm boards placed opposite to each other were selected by visual inspection. Boards with cracks or cell collapse were not used. The edges of the boards were trimmed along the cortex using a table saw. Due to the natural taper of the oil palm trunks (some 0.8 cm per meter trunk length),

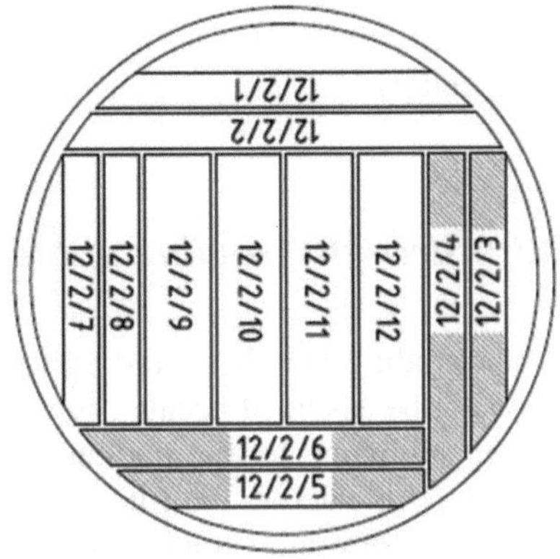

Fig. 1 Sawing pattern of the oil palm trunk sections with board identification numbers

the boards had different widths at each end (approx. 2–3 cm difference). The rough-cut lamellas were calibrated using a two-side planer with HeliPlan tools from Leitz, Germany, one board of each pair to a thickness of 20 mm, the other to 17 mm.

2.2 *Calculation of Lamella Strength Class Limits*

Because there is no existing grading standard for oil palm wood and the small dimensions of the GLT produced within this investigation, the strength grading of the oil palm lumber was based on the European strength class system for coniferous lumber (C-classes) according to [13, 14]. The statistic calculation model of [14], attachment B was used in modified form (Eq. 1) to calculate the density limits for the strength classes, based on the relationship between density as indicating property (IP) and tensile strength ($f_{t,0}$ = 1E-05$\rho^{2.38}$) resp. compression strength ($f_{c,0}$ = 2E-05$\rho^{2.24}$) parallel to the vascular bundles for MOR determined in preliminary investigations on small test specimens, published in [11] and linearized by the natural logarithm. The 5% probability level and therefore $t = 1.645$ was assumed.

$$MOR = a_{MOR} IP + b_{MOR} - t \cdot s_{\delta,MOR} \rightarrow IP = \frac{MOR - b_{MOR} + t \cdot s_{\delta,MOR}}{a_{MOR}} \quad (1)$$

For strength class C14 [13], the characteristic tensile strength value parallel to the fibers ($f_{t,0,k} = 7.2$ MPa) leads to a density limit value of 427 kg/m^3 and the characteristic compression strength ($f_{c,0,k} = 16$ MPa) to a density range of 363–427 kg/m^3. Because of the high share of oil palm wood material with densities below 350 kg/m^3, property values for an assumed strength class C10 were extrapolated with 290–335 kg/m^3 for the compression lamellas and 335–363 kg/m^3 for the tension lamellas.

The density limit value for the shear lamellas in the middle was <290 kg/m^3 for both beam setups.

2.3 *Strength Grading of the Boards*

When grading boards according to their density, the average density of the board must not be assumed because the density gradient over the trunk's cross-section is higher compared to the gradient along the trunk height [15]. Figure 2 shows the density profile across the 140 mm width of a board from the outer area of the trunk.

The 17 mm thick board of each pair was used to "conventional" cut the boards according to their geometry in two 55 mm wide and 2200 mm long lamellas from each side of the board, which resulted in "full size" cross-section (= non-ripped) lamellas (Fig. 2, left). This resulted in lamellas with a density gradient over the cross-section from approximately 800–350 kg/m^3. The mean density of each non-ripped lamella was determined according to DIN 52182 [16].

The 20 mm thick board of each pair was density measured over the board width using X-ray. Therefore, 5 cm long specimens were cut from both ends of the boards, the thickness of the specimens was given by the board thickness. The specimens were planed, rectangular cut and conditioned at a standard climate of 20 °C/65% rh [17]. The X-ray measurements were performed using DENSE-LAB X from Electronic Wood Systems (EWS), Hameln. The specimens were aligned in the holding device in a way that allowed the X-ray beam to travel through the specimen at the middle of the specimen thickness. The specimen holder moved with a constant measuring speed

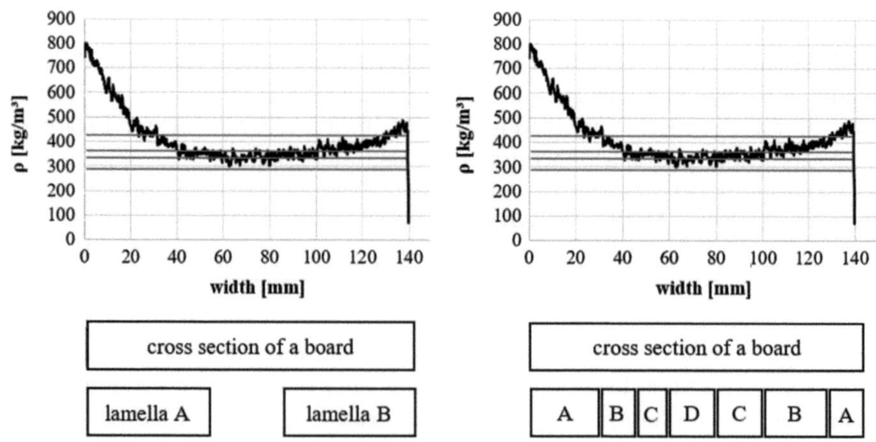

Fig. 2 Density profile of a board from the outer area of an oil palm trunk with a high density gradient over the board width. Cutting of "full size" lamellas according to their geometry with inhomogeneous density (left) and cutting strips with individual width and homogeneous density (right)

of 0.83 mm/s, measurements were taken at every 0.1 mm of the specimen. Based on the measured density profile from the end with the lower density (according to [15], it is assumed that this is the upper end), the cutting positions of each board were calculated from the smoothed density profile with a specially developed software. The minimum strip width is 10 mm. Figure 3 shows an example of a density profile with a smoothed and respectively non-smoothed curve and the calculation of the cutting positions (vertical grey wide columns). The boards were ripped lengthwise from the upper end to strips with individual width (Fig. 2, right). The grade assigned to each strip was verified by determining the mean density of each strip from mass and volume. Due to the longitudinal density distribution, the density is higher at the lower end of the board. This leads to a deviation between the calculated X-ray density class and the measured mean density of the strip. Each strip was marked with the grading class and individual strip number for further traceability. The strips were grouped according to their grade and randomly within the grading class edge-glued into timber boards (Fig. 4). Two to six strips were required to achieve the glued timber board width of 175 mm. A fibre-reinforced, one-component polyurethane adhesive (Jowapur® 686.60) was used for the edge-gluing. After curing and conditioning, the edge-glued timber boards were calibrated to the final thickness of 17 mm (Fig. 4). Three ripped lamellas were cut in width (55 mm) and length (2200 mm) from each edge-glued board. After lamella preparation, the average density was measured in laboratory conditions according to DIN 52182 [16]. Ultimately, the ripped lamellas showed a more uniform density over their width compared to the non-ripped lamellas.

2.4 Production of Glued Laminated Timber (GLT)

Six strength-graded lamellas were arranged within the combined GLT beam according to Fig. 5. Due to the low number of ripped and non-ripped lamellas of grading class <290 kg/m^3, additional non-ripped lamellas were produced from the low density boards from the center of the same trunk sections. The boards were processed in accordance with the manufacturing steps of the non-ripped lamellas described in Sect. 2.3.

The same resin was used for gluing the GLT as for the edge-gluing (Jowapur® 686.60), the mean adhesive application rate was approx. 350 g/m^2. Pressing of the GLT was done at MINDA Industrieanlagen in Minden, Germany on a modified press, TimberPress X 300 (Fig. 6, left), commonly used to produce CLT. The dimensions of the TimberPress X are 3.4 m in length and 1.2 m in width. The press has two lengthwise divided pressing plates (Fig. 6, right), each 0.6 m wide, which can be individually controlled by a hydraulic pressure system. The beams were pressed with the specific pressure of 0.75 MPa for a minimum of 240 min. After storing the beams in a standard climate [17], the beams were calibrated to the final width of 50 mm and stored again in the standardized climate until testing. In summary, 20 beams with different beam setups and from different lamella strength classes were produced.

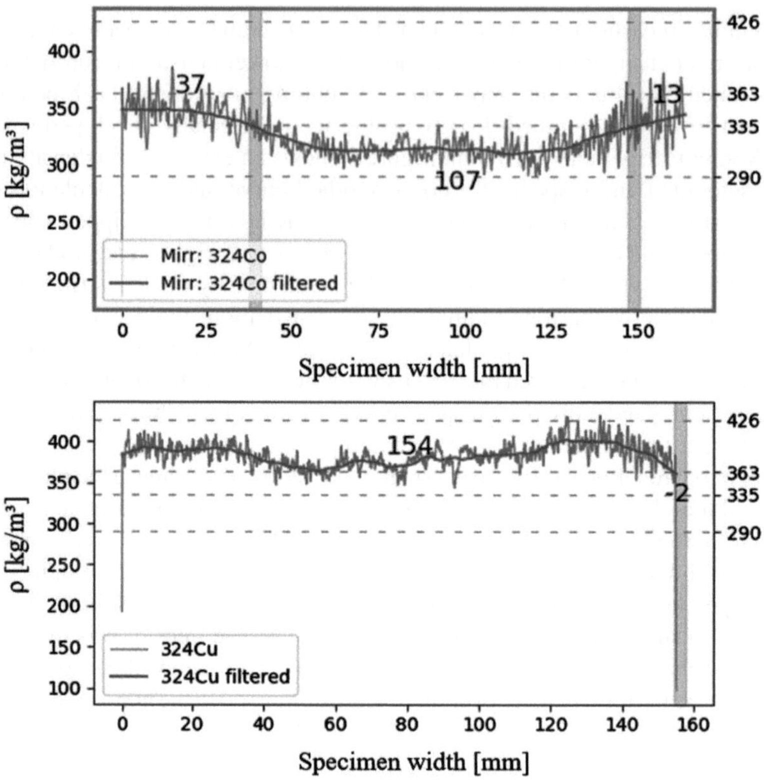

Fig. 3 Calculation of cutting positions and strip width based on the radiometric measured density profile (with and without smoothing) of the specimen from the upper and lower board end of 3/2/4 C

2.5 Testing Methods

The mean density of the beams was determined according to DIN 52182 [16]. Modulus of rupture (MOR) and local and global modulus of elasticity (MOE) of the GLT were determined in a four-point bending test according to [18]. The dimensions of the beams were 50 mm in width, 102 mm in height and 2200 mm in length. Due to expected large deflections, the specimen length corresponds to 21.5 times the specimen height instead of 19 times the height as recommended by EN 408 [18] to prevent the beam from slipping off the supports. A 20 kN testing machine was used for the C10 and a 100 kN testing machine for the C14 beams. For measuring the local and global MOE of the C10 beams, three inductive displacement transducers were used (two of type WA50 and one of type WA100; Hottinger Baldwin Messtechnik (HBM), Germany). The displacement values were recorded via a measuring amplifier (type

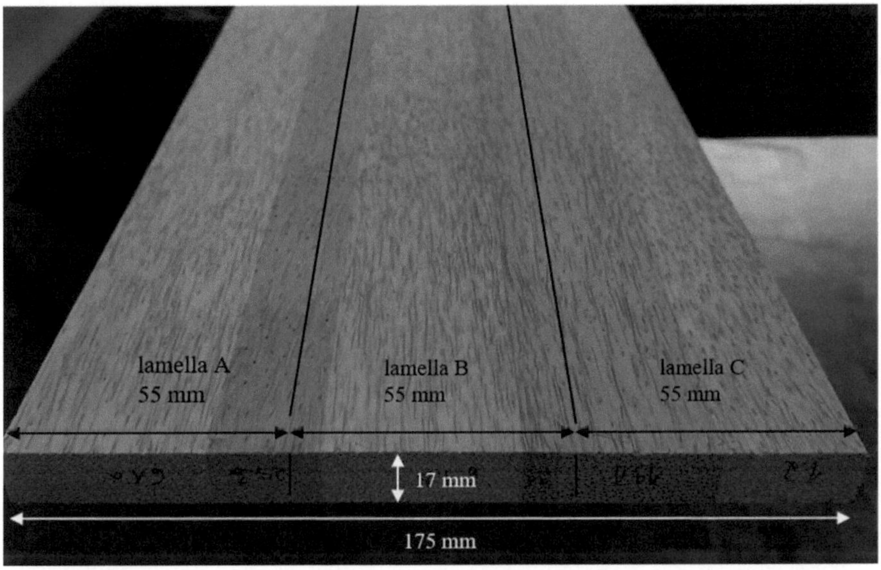

Fig. 4 Cross section of a calibrated edge-glued board for the production of ripped lamella based beams

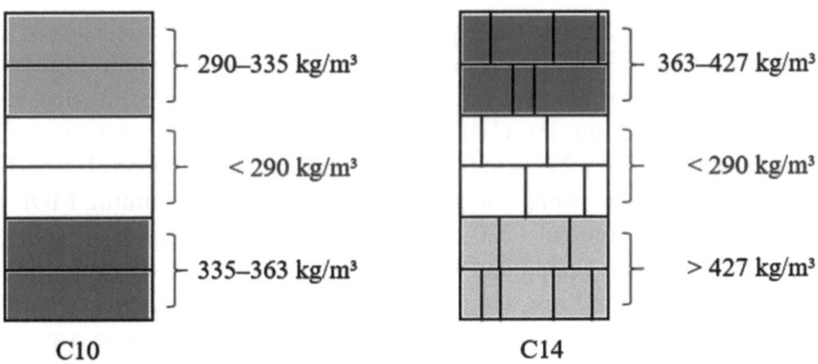

Fig. 5 Combined beam structures to achieve strength class C10 (left) and C14 (right). The left beam is based on non-ripped lamellas, whereas the beam on the right is based on ripped lamellas

HBM MX480A 4-channel measuring amplifier). The data as well as the control and calibration of the displacement transducers was recorded using the MX Assistant 4 software from HBM. The sampling rate of the displacement transducers was adapted to the sampling rate of the universal testing machine (100 Hz). The test speed was chosen so that any beam failure would occur within the time limit specified by EN 408 [18]. To determine the global MOE of the C14 beams a potentiometric displacement transducer was used (type 8712-10, burster Präzisionsmesstechnik, Germany). The local MOE was measured with the same inductive displacement transducers as used for the C10 beams (two of type WA50, HBM). The sensors were directly

Fig. 6 TimberPress X 300 (left) for pressing the GLT. Alignment of the GLT beams in the press, with lateral restraint between the beams (right)

connected to the computer of the testing machine, no measurement data amplifier was required. The test speed was controlled manually via a hydraulic unit. Plywood and oil palm wood supports of various dimensions were used to protect against support roller imprints when testing the C10 beams, plywood supports were used for the C14 beams. Deviating from EN 408 [18], the length of the supports (plywood) were three times the specimen width. The measuring range for determining MOE of the C10 beams was up to a force of 1200 N for both beam setups and for the C14 beams 600…1600 N for the non-ripped and 600…2400 N for the ripped beams.

Specimen geometry and test setup were corrected with k_h and k_l according to EN 384 [19]. The calculation of the characteristic values was done according to EN 14358 [20]. For the characteristic density (ρ_k) and strength ($f_{m,k}$) values, the parametric approach was taken due to the small number of test specimens and under the assumption of a normal distribution. Instead of the reduction factor k_s (n), which depends on the number of specimens, the value of the 5% quantile for a standard normal distribution (p0.05 = 1.645) was used in all calculations of the characteristic values in addition to the calculation according to the normalized reduction value for k_s (n).

3 Results and Discussion

3.1 Fracture Patterns and Failure Description

During the loading of the C10 beams on the 20 kN testing machine, unilateral longitudinal compression was observed in all beams either in front of or behind the supports and beneath the load inducing area. When the maximum longitudinal compressive strength was reached, the vascular bundles buckled, leading to compression failure (Fig. 7, left). However, this ductile compression failure did not lead to a drop in force, so the test load was increased further until the beams failed in tension on the outermost lamella. This tension failure usually occurred below the compressive failure. To prevent compression failure parallel to the vascular bundles, different support materials with various lengths were used.

Because of the observed compression failure next to the supports at the C10 beams, the C14 beams were tested with the same load beam, but on a larger (100 kN) testing machine. To avoid buckling of the vascular bundles and subsequent failure in compression, five beams (two non-ripped and three ripped) were tested under reverse loading (upside down). Lamellas with the highest density were on the top (on the compression side) of the beam, because according to Fruehwald-Koenig and Heister [11] the compression strength parallel to the vascular bundles correlates positively to the density. One reverse tested beam (non-ripped) failed in shear, one (ripped) in compression and three in tension (Fig. 7, right).

Fig. 7 Typical fracture patterns; left: in compression below the load inducing area and in the area of the supports; right: in tension on a reverse loaded beam

Table 1 Density values of the C10 and C14 beams

Strength class	Beam setup	$\rho_{\bar{x}}$ (kg/m^3)	ρ_{σ} (kg/m^3)	cv (%)	$\rho_k^{0.05}$ (kg/m^3)	$\rho_k^{k(s)}$ (kg/m^3)	*n*
C10	Non-ripped	361	5	1	353	346	3
	Ripped	403	13	3	381	361	4
C14	Non-ripped	414	10	2	398	390	8
	Ripped	434	14	3	410	399	5

$\rho_k^{0.05}$ = reduction factor k(s) = p(0.05) = 1.645
$\rho_k^{k(s)}$ = reduction factor k(s) according to EN 14358 [20]

3.2 Density

The mean densities of the ripped beams of both strength classes are higher than that of the non-ripped beams (Table 1). The coefficient of variation is in a similar range and varies between 1 and 3%. The difference in the characteristic density ($\rho_k^{0.05}$) of ripped and non-ripped C14 beams is less than that of the C10 beams.

3.3 Bending Strength

Figure 8 shows the positive trend between density (ρ) and bending strength ($f_{m,0,rf}$) after taking into account the reduction factors k_h and k_l (Fig. 8 and Table 2). The bending strength of the ripped beams ($f_{m,0,rf,r}$) is higher than that of the non-ripped beams at the same density and ranges between 22…27 MPa for the ripped C14 and 17…24.5 MPa for the ripped C10 beams. At a density of 400 kg/m^3, the ripped C10 and C14 beams show almost the same bending strength of 22.5…24 MPa.

The characteristic bending strength ($f_{m,k}^{0.05}$) of the ripped C14 beams is 21 MPa and comparable to that of the ripped C10 beams with 20 MPa (Table 2). The difference between the characteristic bending strength of the non-ripped C10 (11 MPa) and C14 (16 MPa) beams is higher. The characteristic values $f_{m,k}^{k(s)}$ according to [20] are 2…3 MPa lower than the $f_{m,k}^{0.05}$ values. The coefficient of variation ranges between 9…13% for the C14 and 7…8% for the C10 beams.

3.4 Bending Stiffness

Figure 9 shows the local MOE ($E_{m,0,l}$) for the different beam setups and strength classes. The range of the local MOE ($E_{m,0,l}$) of C14, especially the ripped beams, is the highest for all setups and strength classes. The lower values of the C14 are similar to the higher values of C10.

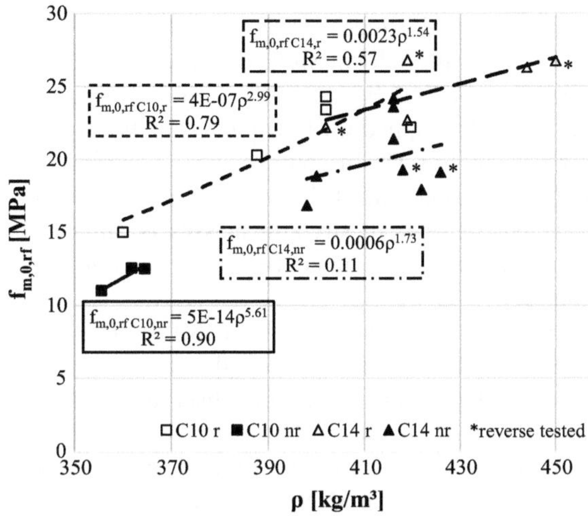

Fig. 8 Relationship between bending strength ($f_{m,0,rf}$) after considering the reduction factors k_h and k_l [19] (EN 384) and the GLT density

Table 2 Bending strength properties of GLT beams class C10 and C14

Strength class	Beam setup	$f_{m,\bar{x}}$ (MPa)	$f_{m,\sigma}$ (MPa)	cv (%)	$f_{m,k}^{0.05}$ (MPa)	$f_{m,k}^{k(s)}$ (MPa)	*n*
C10	Non-ripped	12	1	7	11	9	3
	Ripped	23	2	8	20	17	4
C14	Non-ripped	20	3	13	16	14	8
	Ripped	25	2	9	21	19	5

$f_{m,k}^{0.05}$ = reduction factor k(s) = p(0.05) = 1.645
$f_{m,k}^{k(s)}$ = reduction factor k(s) according to EN 14358 [20]

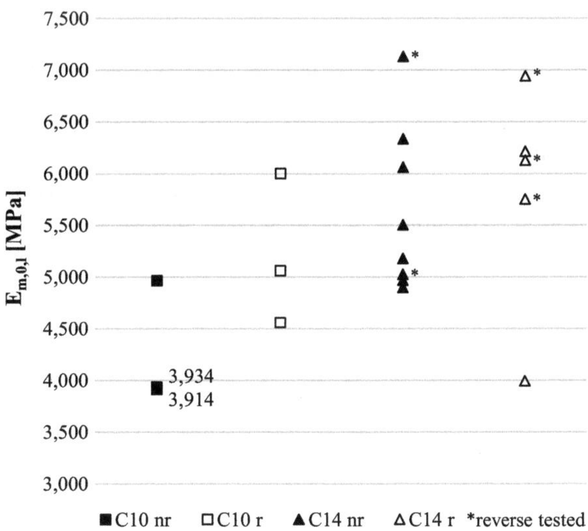

Fig. 9 Local MOE ($E_{m,0,l}$) of the different beam setups and strength classes

The statistical values for the local MOE are shown in Table 3. The ripped beams show a higher mean local MOE ($E_{m,0,1,\bar{x}}$) than the non-ripped beams and the C14 are higher than the C10. $E_{m,0,mean,l}{}^{0.05}$ of the non-ripped C14 beams is higher (4319 MPa) than that of the ripped C14 beams (3993 MPa) (due to one very low MOE value in the C14r, cf. Fig. 9). In contrast, $E_{m,0,mean,l}{}^{0.05}$ of the ripped C10 beams is higher (4000 MPa) than that of the non-ripped C10 beams (3283 MPa). $E_{m,0,mean,l}{}^{0.05}$ for ripped beams of both strength classes is almost similar (3993 MPa for C14 and 4000 MPa for C10). The coefficient of variation is 14%, except for the ripped C14 beams (cv = 19%).

The relationship between local MOE ($E_{m,0,l}$) and MOR ($f_{m,0,rf}$) of beams for different beam setups and strength classes is shown in Fig. 10. A positive trend between stiffness and strength is observed over the entire specimen range. Furthermore, the MOR ($f_{m,0,rf}$) of the ripped beams is higher for both strength classes than of the non-ripped beams. The strength of the ripped C14 beams is in the range of 22…27 MPa at a local MOE of 4000…7000 MPa.

Figure 11 shows the relationship between the local MOE ($E_{m,0,l}$) and global MOE ($E_{m,0,g}$) for oil palm wood beams with different setups and strength classes and the linear relationship for coniferous wood species according to EN 384 [19]. For all specimens, the local MOE ($E_{m,0,l}$) is above the global MOE ($E_{m,0,g}$), which is in accordance with most softwoods. All oil palm wood beams tested showed a positive linear correlation between the global and local MOE. But the linear relationship for softwoods according to EN 384 [19] does not apply to oil palm wood.

4 Conclusion

Density respectively the elastomechanical properties of the lamellas and lamella density-pre-grading and ripping influence the bending properties of GLT made of oil palm wood. Ripping boards according to their density into stripes and edge-gluing the stripes to density homogeneous lamellas results in higher MOR and MOE values for the GLT (due to the positive correlation between density and elastomechanical properties). Even the $f_{m,k}$ for non-ripped C14 beams ($f_{m,k}$ = 14 MPa) is lower than for ripped C10 beams ($f_{m,k}$ = 17 MPa). Designing beams with lamellas placed according to their density was shown to have a significant influence on the elastomechanical properties and fracture pattern of the beams. The local MOE of these beams is higher than the global MOE. The linear regression function between the local and global MOE for coniferous timber according to EN 384 [19] does not apply to beams made from oil palm wood. Due to the low number of specimens and limited variation in lamella densities, it cannot clearly be determined whether the local MOE values of ripped beams are higher than of non-ripped beams.

Table 4 shows the comparison of the characteristic values of strength classes C14–C22 for coniferous timber according to EN 338 [13] and the determined characteristic values for ripped and non-ripped C10 and C14 beams made of oil palm wood. The oil palm wood GLT C10 and C14 achieves the target MOR ($f_{m,k}$). Applying the reduction

Table 3 Local MOE properties of GLT beams class C10 and C14

Strength class	Beam setup	$E_{m,0,l,\bar{x}}$ (MPa)	$E_{m,0,l,\sigma}$ (MPa)	cv (%)	$E_{m,0,mean,l}^{0.05}$ (MPa)	$E_{m,0,mean,l}^{k(s)}$ (MPa)	N
C10	Non-ripped	4271	601	14	3283	3988	3
	Ripped	5205	733	14	4000	4860	4
C14	Non-ripped	5636	801	14	4319	5371	8
	Ripped	5804	1101	19	3993	5440	5

$E_{m,0,mean,l}^{0.05}$ = reduction factor k(s) = p(0.05) = 1.645
$E_{m,0,mean,l}^{k(s)}$ = reduction factor k(s) according to EN 14358 [20]

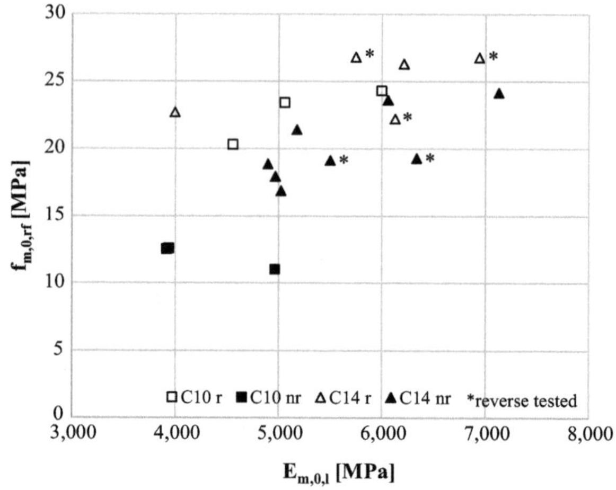

Fig. 10 Relationship between local MOE ($E_{m,0,l}$) and bending strength ($f_{m,0,rf}$) (after considering the factors k_h and k_l [19] (EN 384))

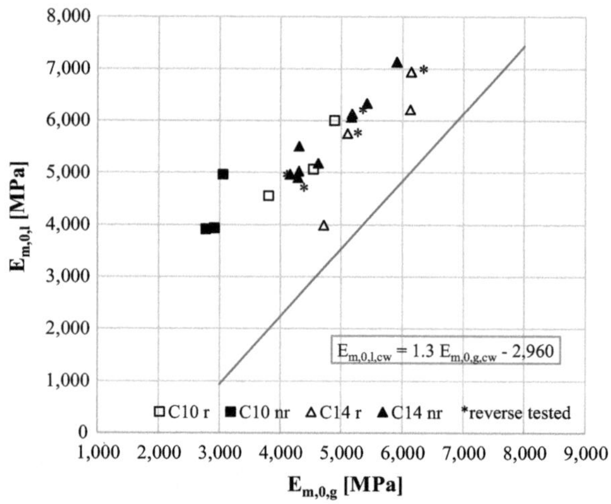

Fig. 11 Relationship between global MOE ($E_{m,0,g}$) and local MOE ($E_{m,0,l}$) for oil palm wood GLT and coniferous wood

factor for coniferous wood according to EN 14358 [20], the target $f_{m,k}$ is not fulfilled by the non-ripped C10 beams. Therefore, the calculated density limits for achieving a target characteristic strength and the calculation method are reasonable in principle. The target MOE for C14 according to EN 338 [13] ($E_{m,0,mean} = 7000$ MPa) is far from achieved. In contrast, the target density ($\rho_k = 290$ kg/m^3) of C14 according to EN 338 [13] is achieved by all oil palm wood beams.

The specific characteristic bending strength ($f_{m,k}/\rho_k$) of the oil palm wood GLT is between 26.0 and 47.6 MPa (Mg m^{-3})$^{-1}$, the ripped CLT are 47.1 (C10) and 47.6 (C14) MPa (Mg m^{-3})$^{-1}$. Therefore, the ripped are in the range of coniferous C14 (48.3 MPa (Mg m^{-3})$^{-1}$). In contrast, the specific mean MOE ($E_{m,0,mean}/\rho_{\overline{x}}$) of oil palm wood GLT is between 11.0 and 13.0 GPa (Mg m^{-3})$^{-1}$ and therefore much lower than that of coniferous C14 (20.0 GPa (Mg m^{-3})$^{-1}$). It can be concluded

Table 4 Characteristic values for strength, stiffness and density of oil palm GLT beams with grade C10[a] and C14[a] and for coniferous timber [13]. The characteristic values for oil palm beams are calculated according to EN 14358 [20]

	Oil palm timber				Coniferous timber				
Strength class	C10[a]		C14[a]		C14	C16	C18	C20	C22
	N-R	R	N-R	R					
$f_{m,k}$ (MPa)	9	17	14	19	14	16	18	20	22
$E_{m,0,mean}$ (MPa)	3988	4860	5371	5440	7000	8000	9000	9500	10000
ρ_k (kg/m^3)	346	361	390	399	290	310	320	330	340
$\rho_{\bar{x}}$ (kg/m^3)	361	403	414	434	350	370	380	400	410

[a]Defined grading classes for oil palm GLT beams according to strength classes for coniferous timber according to EN 338 [13]

that the relationship between the characteristic values for strength and density of coniferous timber according to EN 338 [13] fits for GLT from oil palm, but not the relationship between the mean MOE and density. The MOE is – in relation to strength and density – much lower for oil palm wood.

Acknowledgements The project was funded by the German Federal Ministry of Education and Research through the "Bioökonomie International 2017" project "Oilpalmsugar (031B0767A)". Leitz GmbH & Co. KG provided wood processing tools, Electronic Wood Systems the X-ray measurement device, Jowat SE the adhesives and MINDA Industrieanlagen GmbH the press.

Author Contributions Katja Fruehwald-Koenig is credited for the initial idea of ripping oil palm boards according to their density and the research questions, project goal and research approach. She also developed the oil palm wood strength classes and calculated their density limits. Lena Heister performed all laboratory tests, their analyses, and drew the figures and tables under the supervision of Katja Fruehwald-Koenig. Both authors wrote this paper together.

References

1. Fruehwald A, Fruehwald-Koenig K (2019) The use of oil palm trunks for wood products. In: Materials research proceedings of the 1st world conference on by-products of palm trees and their applications (ByPalma), 15–17 December 2018, Aswan, Egypt, p 11
2. Jumaat MZ, Rahim AHA, Othman J, Midon MS (2006) Strength evaluation of oil palm stem trussed rafters. Constr Build Mater 20(9):812–818
3. Srivaro S, Matan N, Lam F (2019) Performance of cross laminated timber made of oil palm trunk waste for building construction: a pilot study. Eur J Wood Wood Prod 77(3):353–365
4. Killmann W, Lim SC (1985) Anatomy and properties of oil palm stem. In: Proceedings national symposium of oil palm by-products, Kuala Lumpur, Malaysia
5. Lim SC, Gan K (2005) Characteristics and utilization of oil palm stem. Timber Bulletin: Forest Research Institut Malaysia, pp 1–7
6. Balfas J (2006) New approach to oil palm wood utilization for woodworking production part 1: basic properties. Indonesian J Forest Res 3(1):55–65

7. Erwinsyah SH (2008) Improvement of oil palm wood properties using bioresin. Dissertation, Technische Universität Dresden, Institut für Forstnutzung und Forsttechnik, Fakultät für Forst-, Geo- und Hydrowissenschaften, Dresden, 171p
8. Fathi L (2014) Structural and mechanical properties of the wood from coconut palms, oil palms and date palms. Dissertation, University of Hamburg, Zentrum Holzwirtschaft, Hamburg, 159p
9. Srivaro S, Matan N, Lam F (2018) Property gradients in oil palm trunk (Elaeis guineensis). J Wood Sci 64(6):709–719
10. Fruehwald-Koenig K (2019) Mechanical properties of oil palm wood. In: Presentation + Abstract on the CompWood2019, international conference on computational methods in wood mechanics—from material properties to timber structures ECCOMAS thematic conference, 17 June 2019, Växjö, Schweden
11. Fruehwald-Koenig K, Heister L (2022) Macromechanical and micromechanical behavior of oil palm wood (*Elaeis guineensis* JACQ.)—Part 1: tensile, compression and bending properties. Publication in preparation
12. Fruehwald-Koenig K (2019) Properties and grading of oil palm lumber. In: Wang X, Sauter U, Ross RJ (eds) 21th international nondestructive testing and evaluation of wood symposium; Freiburg, Germany, 24–27 September 2019, U.S. Department of Agriculture, Forest Service, Forest Products Laboratory, Madison, WI, pp 204–212
13. EN 338. Structural timber—Strength classes. Beuth Verlag, Berlin, 13p
14. EN 14081-2 (2018) Timber structures—strength graded structural timber with rectangular cross section—Part 2: machine grading; additional requirements for type testing. Beuth Verlag, Berlin, 39p
15. Koelli N (2016) Density and moisture distribution in oil palm trunks from Peninsular Malaysia. B.Sc thesis, University of Hamburg, Zentrum Holzwirtschaft, Hamburg, 52p
16. DIN 52182 (1976) Testing of wood; determination of density. Beuth Verlag, Berlin, 3p
17. DIN 50014 (2018) Standard atmospheres for conditioning and/or testing—specifications. Beuth Verlag, Berlin, 7p
18. EN 408 (2012) Timber structures—structural timber and glued laminated timber—determination of some physical and mechanical properties. Beuth Verlag, Berlin, 39p
19. EN 384. Structural timber—determination of characteristic values of mechanical properties and density. Beuth Verlag, Berlin, 21p
20. EN 14358. Timber structures—calculation and verification of characteristic values. Beuth Verlag, Berlin, 17p

Effects of Density and Resin Content on Particleboard from Oil Palm Frond (OPF)

Nurrohana Ahmad, Ainul Munirah Abdul Jalil, Zaimatul Aqmar Abdullah, Siti Noorbaini Sarmin, and Ahmad Naqi Razali

Abstract Oil palm frond (OPF) can act as an alternative source for wood-based industry. OPF was an agricultural waste, and it is producing millions of tons of oil palm biomass every year. The oil palm frond was obtained from oil palm plantation UiTM Pahang. The frond part of the oil palm was separated for further processes. The urea–formaldehyde (UF) resin was used as a binder. The particleboard samples were then tested for their mechanical and dimensional stability of OPF by using different resin content (8, 10, and 12%) and board density (500, 600, and 700 kgm^{-3}). The tests on the mechanical properties include Modulus of Elasticity (MOE), Modulus of Rupture (MOR) and Internal Bonding (IB). The dimensional stability tests are thickness swelling (TS) and water absorption (WA). The results showed that the higher density board and resin content seem to have a better result in mechanical properties and dimensional stability. Statistical analysis indicated there was significant differences between boards made from different resin content except for thickness swelling (TS). The TS value exceeded the maximum set value (12%), but highly significant differences were observed in the density. Boards with a density of 700 kgm^{-3} and 12% of resin content achieved the minimum requirement based on the JIS A5908:2003 standard type 8 for mechanical properties. Overall, the particleboard from oil palm fronds is suitable to be used as an alternative resource to overcome the shortage of raw material in the furniture and wood industry.

N. Ahmad (✉) · A. M. A. Jalil · Z. A. Abdullah · S. N. Sarmin
Department of Wood Technology, Universiti Teknologi MARA Pahang, Bandar Jengka, Malaysia
e-mail: nurrohana@uitm.edu.my

A. M. A. Jalil
e-mail: ainulmunirah@uitm.edu.my

Z. A. Abdullah
e-mail: zaimatul@uitm.edu.my

S. N. Sarmin
e-mail: baini@uitm.edu.my

A. N. Razali
Veolia Water Technologies Asia Pacific, Kuala Lumpur, Malaysia

M. Jawaid et al. (eds.), *Proceedings of 2nd World Conference on Byproducts of Palms and Their Applications*, Springer Proceedings in Materials 19,
https://doi.org/10.1007/978-981-19-6195-3_4

Keywords Bending strength · Dimensional stability · Urea-formaldehyde

1 Introduction

The utilization of particleboard is one innovative way to use oil palm biomass. It is expected that particleboard from oil palm. Oil palm frond (OPF) has a very high potential to be developed as an alternative resource for wood-based products after all the conservative wood products such as rubberwood, acacia, and others. Apart from being tagged as biomass or waste product, there are also efforts to make OPF as the basic material for the pulp and paper industry, feedstocks, composite board [1], and also as a filler in concrete [2]. This shows that oil palm frond is kind of a flexible material, commercialized, and can be turned into value-added products in the wood-based industry and commands a high market prospect. The oil palm fronds can be used to produce quality composite boards that possess the physical and mechanical strength values that are better than the composite boards made from oil palm stem and less than rubberwood [3].

OPF are currently considered waste from oil palm plantations, and their biomass is not used completely. Production from these plantations by pruning and replanting programs generated about 18 million tons of fronds a year all over the world. The oil palm frond is considered one of the new sustainable lignocellulosic raw materials if we use it effectively [4]. A product can substitute the standard particleboard currently used. OPF particleboard may encourage better utilization of total biomass produced and the most viable alternative to be utilized as value-added products. In some cases, some manufacturers have not used these resources effectively. Open burning and landfilling are common ways for them to eliminate the oil palm residue which later cause pollution issues to the environment [5].

Among the generated oil palm biomass, OPF is by far the highest in quantity, accounting for more than 50% of the total biomass generated in the oil palm industries. Therefore, OPF appears to be a very promising source of raw materials in Malaysia. OPF has the potential to serve as a sustainable and promising biomass source for Malaysia, not only to reduce the environmental problem caused by the openly burnt or landfill disposal of OPF but also to generate extra revenue for the country [6].

The objective of this study was to investigate the properties of OPF particleboard with urea–formaldehyde as binders focusing on the effect of resin content (8, 10, and 12%) and density (500, 600, and 700 kgm^{-3}). In addition, the mechanical and dimensional stability of the particleboard were determined.

2 Materials and Methods

Materials. OPF was collected from the Oil Palm Plantation UiTM, which was intentionally planted for academic purposes. The frond part of the oil palm was separated for further processes. The palm frond was converted into a flake and chip structure using a knife ring flaker machine and chipper machine to produce a small particle of many sizes. The particles then undergo air drying before screening to prevent insect and fungi attacks until reach equilibrium moisture content (EMC) around 14%. The screening process was done by using a Gilson screener machine. The screener machine was used to separate the dust and the particles from the oil palm frond have quite a fibrous characteristic. After acquiring the particles that have been dust-filtered, the particles need to be dried once again to ensure the moisture content was below 5%. This was due to a possibility that particles were being exposed to the environment (at the idle state after screening) there might be water molecules that are absorbed by the frond particles.

Methods. After that, the particle was mixing using a particleboard mixer with urea–formaldehyde and a hardener. Then, the mixture was then formed in a mat-forming with dimensions 340 mm × 340 mm. The particles were then sent to the cold press part to get the early shape of the board for 30 s. A wooden block was placed on top of the cold press part. A cold-press is important, it to make the board thinner. This can reduce the possibility of the board blowing while being hot-pressed. It is also being done to increase board density and strength. Then the steel bar was placed on the side of the mat with 12 mm thickness. As for the hot press, it is done to cure the adhesive and to reduce the board to the final thickness. The time that it took to cure was about 6 min. The temperature for the hot press used was around 165° C (Fig. 1).

The boards were then trimmed so that a more stable board composition was obtained even on the edge side. It was then being cut into standard size according to the bending size, internal bonding size, thickness swelling size and water absorption

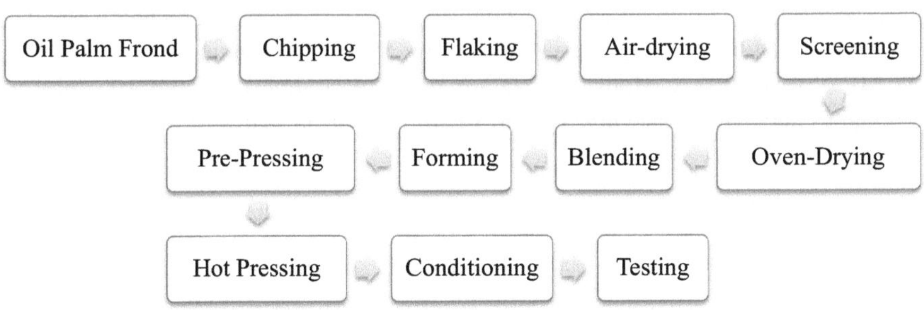

Fig. 1 Particleboard manufacturing flowchart

size. After that, each sample size was measured in terms of thickness, length, width, and weight. The electronic balance machine was used to measure the weight of the sample in grams. Meanwhile, width, length and thickness were measured by using a micrometre calliper and veneer calliper. The average data was calculated and recorded for each sample.

Testing. The sample was test to determine the bending strength and dimensional stability of the particleboard. For the mechanical test, the tests were bending (modulus of rupture and modulus of elasticity) and internal bonding tests and for the dimensional stability testing, the tests conducted were thickness swelling and water absorption.

3 Results and Discussions

These oil palm biomass especially OPF, are the most abundant renewable resources for alternative bioenergy, biomaterials, and wood-based industries [7]. The investigation result on wood-based industries showed a possibility of producing oil palm frond to particleboard with acceptable strength with short rotation also reduces waiting time for a raw material resource.

Table 1 shows the analysis of variance (ANOVA) on mechanical properties and dimensional stability of particleboard made from OPF. There were highly significant differences for resin content on the mechanical properties and dimensional stability of the OPF except for the TS. Moreover, the density was highly affected for the mechanical and dimensional stability in the OPF. It shows highly significant difference between densities in boards made from OPF.

Effect of Resin Content. Figure 2 shows the effect of resin content on the mechanical properties and dimensional stability. There were different strengths recorded on each reading of 8, 10, and 12% of resin content. The highest value of MOR, MOE, and IB was recorded by 12% of resin (12.95, 2051, and 0.26 MPa), respectively. The lowest MOR, MOE, and IB were recorded by 8% resin content (7.23, 1331, and 0.08 MPa), respectively. Both two classes of resin content of 8 and 10% are not significant except for water absorption. Resin content is believed to have a strong

Table 1 The ANOVA on mechanical properties and dimensional stability of particleboard from OPF

SOV	df	MOR (MPa)	MOE (MPa)	IB (MPa)	WA (%)	TS (%)
Resin Content (RC)	2	30.36**	17.86**	11.53**	51.59**	0.72 ns
Density (D)	2	137.18**	86.16**	33.08**	19.51**	15.98**
RC x D	4	10.39 ns	2.45 ns	9.25**	6.38**	0.88 ns

Note SOV—Source of Variance, *df*—Degree of Freedom, ns—not significant $P > 0.05$, *significant at $p < 0.05$, **highly significant at $p < 0.01$

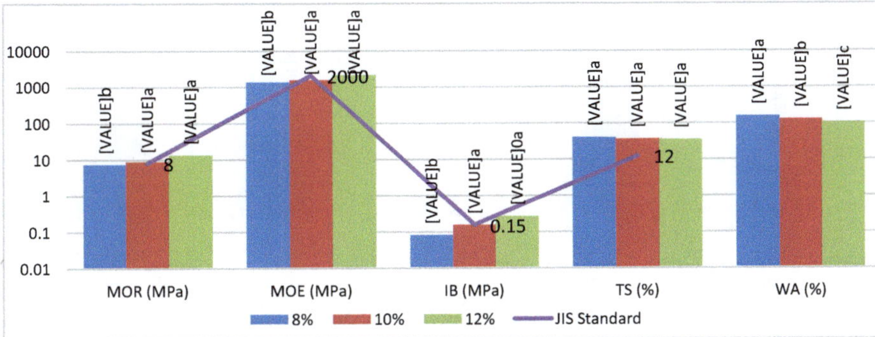

Fig. 2 Effect of resin content on the mechanical and dimensional stability of OPF

influence on the strength properties of particleboard except for thickness swelling. Resin content at 12% were pass the minimum requirement JIS A5908:2003 standard for Type 8 (base particleboard) for the mechanical properties. Based on previous studies, the higher the resin content, the stronger the particleboard [8]. According to Kimoto et al. [9], with increasing the amount of the resin content, there was improved the bending strength value of the board.

Effect of Density. Figure 3 shows the effect of density on the bending properties and dimensional stability of OPF particleboard. There was different strength recorded on each reading of 500, 600 and 700 kgm^{-3} of board density. The highest value for all properties was recorded by 700 kgm^{-3} of density with values (16.58, 2562, 0.15 MPa, 23.09%, and 107.31%), respectively. Meanwhile, the lowest reading was recorded by the density of 500 kgm^{-3}. There are highly significant differences between all three densities except for dimensional stability. It is shows that as the density increases, the value of mechanical properties and dimensional stability also increases. The density of the board is influenced by the value of particles used. Board from 700 kgm^{-3} density was met the minimum requirement for the JIS A5908:2003. Type 8 for mechanical properties. A bigger value of particles will reduce voids between particles hence enhance the resistance to rupture [8]. According to Enayati and Eslah [10], the density of the boards is one of the main factors affecting the properties of particle boards and other wood composites, and an increase in density leads to improved board properties.

4 Summary

Briefly, the OPF product can be utilized for further development based on their strength properties and dimensional properties especially in the making of furniture-based material such as particleboard. This can reduce the tension on current furniture sources and as well can help manage waste properly. In terms of production cost,

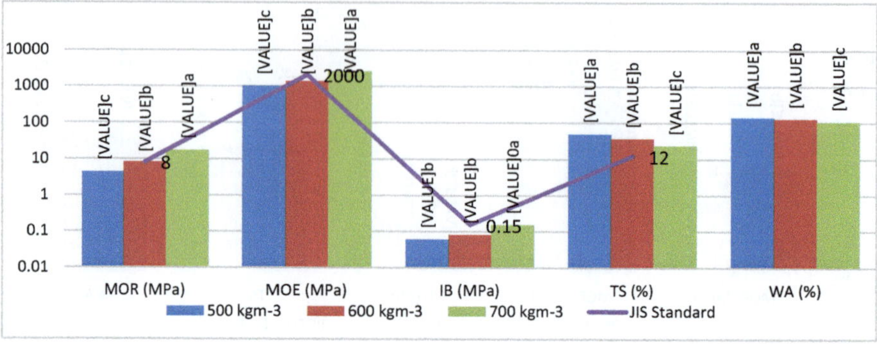

Fig. 3 Effect of density for mechanical and dimensional stability of OPF

it can reduce the raw material cost for furniture as the material is just considered as only a biomass product before. Moreover, the quantity of the untreated biomass from oil palms also strengthening the cause for this material to be commercialized further.

Regarding the result, it was found that the board either with the highest density or the highest resin content is the only class from the experiment that achieved the standard of mechanical properties based on the JIS A5908 standard Type 8. The standard was achieved through tests of MOE, MOR, and IB. However, for the dimensional stability test; TS, and WA of all boards did not achieve the minimum requirement set by the standard. As the density and the resin content increase, the board will have better mechanical and dimensional stability. For resin content, the resin 12% and density of 700 kgm^{-3} passed the mechanical properties standard set by the JIS and it is the best board compared to other boards with lower resin or lower density.

As for the recommendation in the future, there is a possibility to ensure the board to be industrially fit by using a hybrid technique. The board can be mixed with other materials or even other biomasses to ensure the board has better physical properties but further research is required.

Acknowledgements The authors would like to acknowledge UiTM Kampus Jengka, Pahang for the fronds supplies and Malaysian Adhesives Chemical (MAC) Shah Alam for providing us urea-formaldehyde resins for this research.

References

1. Iling E, Siti D, Ali H, Osman MS (2019) Effect of pressing pressure on physical and mechanical properties of elaeis guineensis fronds composite board. e-Bangi 16(3):1–12
2. Yassin MH, Lakys R, Ahmed T, Al-Refaei S, Omar BA-S, Altaher RS (2020) Optimizing the thermal resistance of concrete using the palm tree fronds fibers. No. Cic, pp 452–461. https://doi.org/10.29117/cic.2020.0057

3. Mohd Sukhairi MR, Razak W, Othman, Sulaiman Janshah AT, Aminuddin M, Tamer M, Izyan K (2011) Properties of composite boards from oil palm frond agriculture waste. Boards 6:4389–4403
4. Laemsak N, Okuma M (1998) Development of boards made from oil palm frond II: properties of binderless boards from steam-exploded fibers of oil palm frond. J Wood Sci 46:322–326. https://doi.org/10.1007/BF00766224
5. Kamarulzaman N, Mohd Ariff J, Mansur A, Hashim WS, Abdul Hamid S, Zaihan J (2004) Minimizing the environmental burden of oil palm trunk residues through the development of laminated veneer lumber products. Manag Environ Qual 15(5):484–490. https://doi.org/10.1108/14777830410553924
6. Ooi ZX, Teoh YP, Kunasundari B, Shuit SH (2017) Oil palm frond as a sustainable and promising biomass source in Malaysia: a review. Environ Process Sustain Energy 36(6):1864–1874
7. Financie R, Moniruzzaman M, Uemura Y (2016) Enhanced enzymatic delignification of oil palm biomass with ionic liquid pretreatment. Biochem Eng J 110:1–7. https://doi.org/10.1016/j.bej.2016.02.008
8. Mohamad Jani S, Izran K (2012) Mechanical and physical properties of low density kenaf core particleboards bonded with different resins. J Sci Technol 4(1):17–32
9. Kimoto K, Ishimori E, Sasaki H, Maku T (1964) Effects of resin content and particle dimension on the physical and mechanical properties of low-density particle boards. http://hdl.handle.net/2433/52927
10. Enayati AA, Eslah F (2014) Modeling beech (Fagus orientalis) particleboard properties based on resin content and board density. J Indian Acad Wood Sci 11(1):45–49. https://doi.org/10.1007/s13196-014-0116-0

The Impact Response of Coconut Fibreboards with Corn Starch (CS), Tapioca Starch (TS) and Rice Flour (RF) as Natural Binders

Noor Leha Abdul Rahman, Aidah Jumahat, Napisah Sapiai, and Syed Tajul Muluk Syed Ahmad Putra

Abstract Coconut palm is mainly cultivated in coastal areas of tropical countries. The husk is abundantly available in many places as cheap residue from coconut processing, which is considered to produce coconut fiber. This coconut fiber is explored for new usage and adopted for various engineering applications. It is expected that this fiber would act as an excellent raw material for panel product such as fiberboard with value added. This paper investigates the impact response of fiberboard made from coconut fiber. Three types of natural binder were used which included corn starch (CS), tapioca starch (TS), and rice flour (RF). The coconut fiberboards (CFB) were coded as CFBCS, CFBTS, and CFBRF based natural binder which were prepared using hot compression machine with the temperature of 150 °C for 1 h holding time. The impact behavior of coconut fiberboard was investigated using a drop-weight impact tower at three different energy levels of 5, 10, and 15 J as according to ASTM D7136. The results showed that CFBCS has higher energy absorbed and impact strength followed by CFBTS and CFBRF. For instant, at 15 J of impact energy, the CFBCS possesses 1.51 J energy absorbed and 11.54 kJ/m^2 impact strength, which is 128.80% and 132.66% higher compared to CFBRF, respectively. Among all the natural binders, the corn starch is the best binder due to hydroxyl groups contain in corn starch, which form strong bonds with the coconut fibers. Thus, eventually enhanced impact properties of CFBRF. The optical analysis showed that the coconut fiberboard cracked and chirped during the impact test. Therefore, the formulation of coconut fiber with corn starch could be used for fiberboard in furniture industries.

N. L. A. Rahman (✉) · A. Jumahat · N. Sapiai · S. T. M. S. A. Putra
School of Mechanical Engineering, College of Engineering, Universiti Teknologi Mara, 40450 Shah Alam, Selangor, Malaysia
e-mail: noorleha3585@uitm.edu.my

A. Jumahat
Institute for Infrastructure Engineering Sustainable and Management (IIESM), Universiti Teknologi MARA, 40450 Shah Alam , Selangor, Malaysia

M. Jawaid et al. (eds.), *Proceedings of 2nd World Conference on Byproducts of Palms and Their Applications*, Springer Proceedings in Materials 19,
https://doi.org/10.1007/978-981-19-6195-3_5

1 Introduction

Fiberboard is one of the engineer base products made from wood or other fiber. There are three types of fiberboard which are low density fiberboard (LDF), medium density fiberboard (MDF), and high density fiberboard (HDF). Wood and sawdust are usually used as raw materials for fiberboard. Fiberboard is often known as the medium density fiberboard (MDF) for furniture industry which could be applied in automotive and construction industries. In 1960, the first MDF was developed at the United States. After the first development, the demand for fiberboard increased from year to year. The main source of fiberboard is natural wood which has some weaknesses like very expensive, difficult to process in various shapes, takes a long time to grow and might deplete if continuously used [6, 13]. Therefore, there are several methods to reduce the consumption of wood in making fiberboard. One of the methods currently developed is using natural and waste fibers such as waste poultry feather, kenaf fiber, coconut fiber, hemp, and others as alternative materials to replace the usage of wood [2, 4, 8, 10, 11]. Among all the potential natural fibers, coconut fiber is recognized as a potential material for manufacture fiberboard since the mechanical strength, swelling and thermal conductivity properties have fulfilled the requirements of Japanese JIS A 5905 (2003) standard for fiberboards [4]. Coconut fiber, commonly known as coir, is obtained from the fibrous husk of the coconut. The coconut fruit, which is cultivated for milk and pulp, has leftover coconut shell or crust that limits the use and normally discarded as garbage. The coconut fiber has the highest content of lignin with up to 30%, which makes it harder, tougher and stiffer. Thus, it provides higher compressive strength of cell fiber to protect the carbohydrates from chemical and physical damage [8]. Garcia et al. developed the milled coconut husk and bonded it with tannin-based adhesives (Coconut Fiberboard-CFB) to study the resistance of CFB against insects and fungal attack. The results found that both low and medium density of CFB were highly resistant to subterranean and dry wood termites after 12 months of exposition. Low density CFB showed 11.1% of termite attack while medium density CFB produced 52.3% of termite attack. Ngadiman et al. [8] claimed that the coconut fibreboard with urea gam produced stronger, long lasting, less swelling, low water absorption, high modulus of rupture and low modulus of elasticity as compared to conventional panel board. Furthermore, the study of coconut fiber with epoxy resin by Alotaibi et al. [3] reported that flexural and compressive strengths increased with the increase of coconut fiber volume fraction.

Most of the formulation of MDF use formaldehyde-based resin as adhesive such as phenol formaldehyde, urea formaldehyde and others. Even though formaldehyde-based resin has a lot of advantages such as excellent adhesion properties, water resistance, ease of handling, low curing temperature, short press times, and relative cost-effectiveness, the formaldehyde has been considered a dangerous toxic chemical due to public health concern for the past 30 years [1, 5, 9, 10]. Prolonged exposure of formaldehyde may cause effects on human health such as chronic toxicity, nasopharyngeal cancer, myeloid leukemia, nervous system irritation, skin sensitization, nausea, and lymphohematopoietic malignancies. International Agency for Research on Cancer classified formaldehyde as "carcinogenic to humans", which led to the adoption of stricter legislative regulation on formaldehyde emission values. This serious problem has made the researchers seek alternatives to replace the use of formaldehyde-based resin with less toxic, eco-friendly, renewable and bio-based adhesives. The bio-based adhesives such as starch, sugar, cellulose, plant oils and protein are known as natural binders which have been extensively studied and reported by numerous researchers [1, 4, 7, 15]. For example, Andrew et al. studied the use of starch in MDF to replace urea–formaldehyde. They concluded that biodegradable starch-based produces materials with comparable strength (10–20 MPa) and ductility significantly higher (typically 0.5%.) than those produced with formaldehyde-based resins [1].

The aim of this research work is to investigate the potential of different source of natural binder in producing eco-friendly fiberboard from coconut fibers. Three different types of natural binder, i.e., corn starch (CS), tapioca starch (TS), and rice flour (RF) which are easily found in convenience store, were used to fabricate coconut fiberboard. The study focused on the capability of coconut fiberboard to restrain the damage under impact loading.

2 Materials and Methods

Coconut fiber was collected from a farm in Malaysia. Coconut fiber was crushed using crusher machine to get the length of <5 mm. Three types of natural binders of corn starch (CS), tapioca starch (TS), and rice flour (RF) were used. All natural binders were easily found and purchased from convenience store. The nutrition information is summarized in Table 1.

The fabrication of coconut fiberboard started with the crushing and drying of the coconut fiber. Then, the coconut fiber and natural binder were weighed with the ratio of 70:30 (coconut fiber: natural binder). The natural binder was then diluted with water before homogeneously mixed with coconut fiber. The mixture was placed in 50 mm × 50 mm mould plate and pressed at temperature of 200 °C for 1 h using hot pressing machine. The specimen fabrication process of coconut fiberboard is illustrated in Fig. 1.

Table 1 Nutrition information of corn starch, tapioca starch and rice flour

Nutrition information (serving size: 100 g)	Corn starch	Tapioca starch	Rice flour
Energy, kcal	364	346	356
Carbohydrate, g	90.3	85.3	80.6
Protein, g	0.1	0.10	7.4
Fat, g	0.15	0.30	0.47

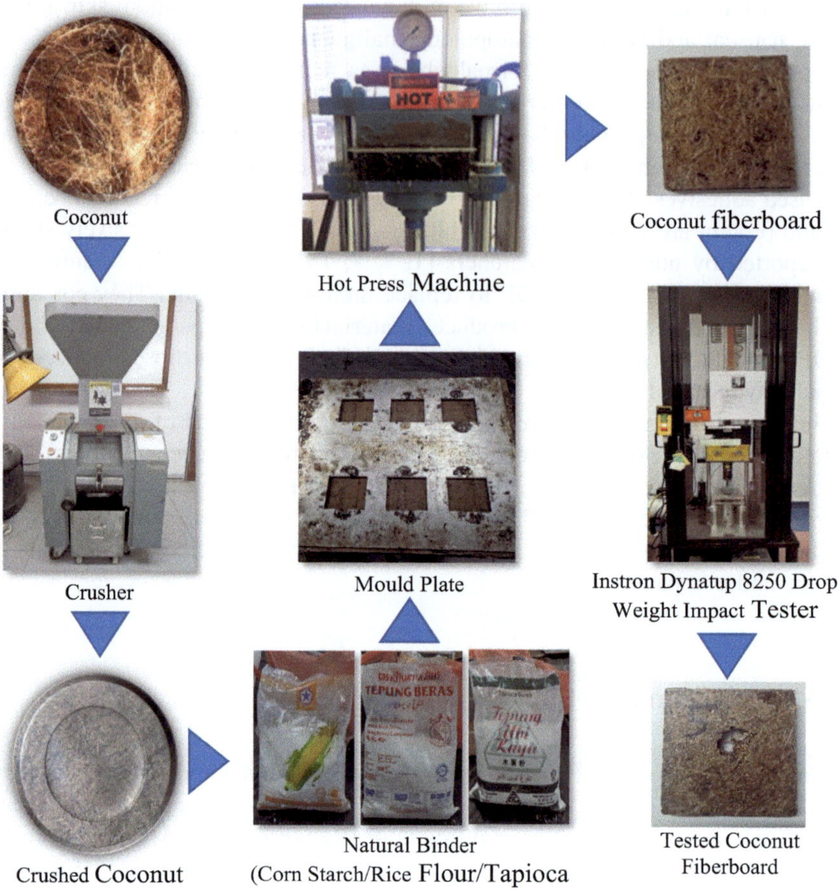

Fig. 1 Fabrication process of coconut fiberboard (CFB)

The impact properties were measured using instrumented Instron Dynatup 8250 drop weight tower according to the standard ASTM D3763. The specimens were tested under three different impact energy of 5, 10, and 15 J using 13 mm diameter of hemispherical tip impactor. The impact properties of load carrying capabilities, energy absorbed, deflection, damage, and impact strength of specimens were measured. Five specimens for each different binder coconut fiberboard were tested. The damage mechanism of impacted specimens was then observed under digital camera.

3 Results and Discussion

The low velocity impact test of CFB was carried out with three different impact energy of 5, 10, and 15 J. The impact properties of peak load, deflection at peak load, total energy absorbed, energy to maximum load and impact strength of three different CFBs, i.e., CFBCS, CFBTS, and CFBRF, were analyzed and summarized in Table 2 and Fig. 2. In Table 2, as expected, the peak load and energy absorbed for all CFBs increased as impact energy level was increased. CFBCS, which used corn starch, had a good impact performance as compared to tapioca starch (CFBTS) and rice flour (CFBRF). At 15 J of impact energy for CFBCS, the peak load and energy absorbed achieved about 703.5 N and 1.51 J respectively, which presented the highest impact performance. Meanwhile, CFBRF had the lowest impact performance as the peak load and energy absorbed recorded about 654.7 N and 0.66 J. Figure 2 illustrates that the impact strength for all types of CFB increased with the increase of impact energy. The impact strength refers to a material's resistance towards breakage. It is also related to the toughness properties of a material. For CFBCS, the impact strength was recorded at about 5.19 kJ/m^2, 6.62 kJ/m^2, and 11.54 kJ/m^2 at impact energy of 5 J, 10 J, and 15 J, respectively. Meanwhile, the impact strength of CFBTS was recorded at about 4.44 kJ/m^2, 7.37 kJ/m^2, and 10.45 kJ/m^2 at impact energy of 5 J, 10 J, and 15 J, respectively. As for CFBRF, the impact strength did not give big changes even at higher impact energy level (15 J). The CFBC produced the highest impact strength among the CFBs especially at high impact energy level (15 J) followed by CFBTS and CFBRF. This finding was supported by the surface damage images (Tables 3, 4 and 5) since CFBRF had a large area of fiber detached which indicated poor adhesion bonding between coconut fiber and rice flour. Besides that, rice flour had the lowest carbohydrate content of about 80.6 g compared to corn starch with 90.3 g and tapioca starch of 85.3 g at the size serving of 100 g. Chemically, starch is a carbohydrate polymer consisting of anhydroglucose units linked by α-D-(1, 4) glucosidic bonds [7, 12]. Starch contains many hydroxyl groups which can wet the polar surfaces of cellulose, penetrate crevices and pores; thus, form strong bonds [5, 14, 16]. Therefore, in this study, the amount of carbohydrate in natural binder may influence the adhesion bonding, thus affected the mechanical properties of CFB.

Tables 3, 4, and 5 show the surface damage images of CFBCS, CFBTS, and CFBRF, respectively after being subjected to impact loading. The images present the impact damage from front and back views of CFB specimens. From the observation, all the CFBs showed the impact damage holes and cracks with fiber breakage. The primary damage modes are indentation induced in the surface, which the indenter penetrated and perforated into CFB specimens. The fracture mode of this fiberboard is mainly characterized by fiber pullout, with smooth surface inside the remaining holes, which indicates poor adhesion between the natural binder and coconut fibers. From the back view, there is obvious observation of the fiber being separated with the binder. The biggest area of fiber detached indicates the weakest adhesive bonding

Table 2 Impact properties of coconut fiberboard (a) CFBCS (b) CFBTS, and (c) CFBRP

(a)	CFBCS			
Impact energy (J)	Peak load (N)	Deflection at peak load (mm)	Total energy absorbed, E_t (J)	Energy to max load, (J)
5	270.3 ± 22.05	34.14 ± 1.27	0.69 ± 0.33	0.69 ± 0.33
10	299.2 ± 64.53	28.55 ± 0.79	0.90 ± 0.35	0.88 ± 0.35
15	703.5 ± 24.70	29.19 ± 1.24	1.51 ± 0.46	1.50 ± 0.46
(b)	CFBTS			
Impact energy (J)	Peak load (N)	Deflection at peak load (mm)	Total energy absorbed, E_t (J)	Energy to max load, (J)
5	246.6 ± 11.73	32.28 ± 0.78	0.60 ± 0.15	0.59 ± 0.15
10	549.9 ± 72.33	29.29 ± 0.56	0.99 ± 0.43	0.98 ± 0.43
15	642.80 ± 47.90	29.27 ± 0.54	1.40 ± 0.33	1.39 ± 0.33
(c)	CFBRF			
Impact energy (J)	Peak load (N)	Deflection at peak load (mm)	Total energy absorbed, E_t (J)	Energy to max load, (J)
5	245.31 ± 8.48	33.46 ± 0.20	0.52 ± 0.68	0.52 ± 0.68
10	335.10 ± 92.22	30.32 ± 0.25	0.64 ± 0.21	0.63 ± 0.21
15	654.7 ± 77.06	29.91 ± 0.24	0.66 ± 0.53	0.66 ± 0.53

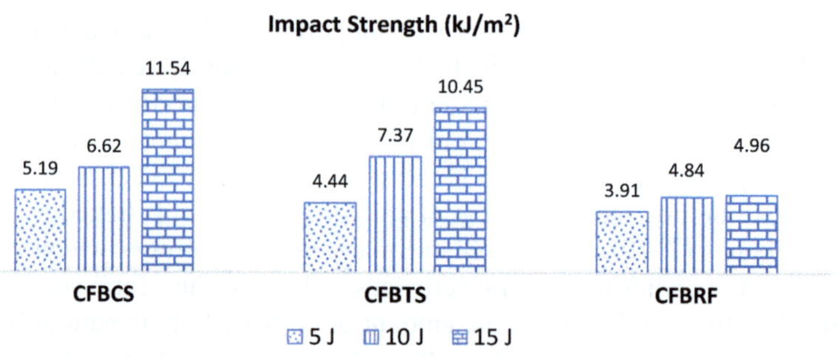

Fig. 2 Impact strength of CFBCS, CFBTS and CFBRP

of natural binder to coconut fiber. In all probability, the fiber has been covered with a thin layer by the binder, thus give a better stress transfer and superior impact strength. In this study, the damage area of CFBRF was larger compared to CFBCS and CFBTS since CFBRF had the lowest energy absorbed and impact strength.

Table 3 Surface damage of CFBCS after subjected to impact test at different impact energy

CFBCS	Impact energy (J)		
	5	10	15
Front view	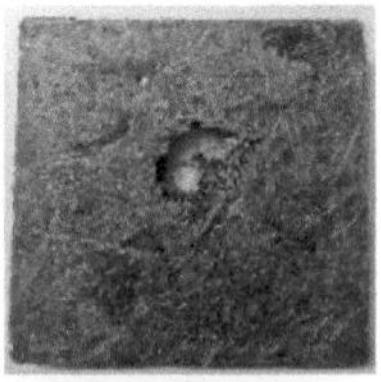		
Back view			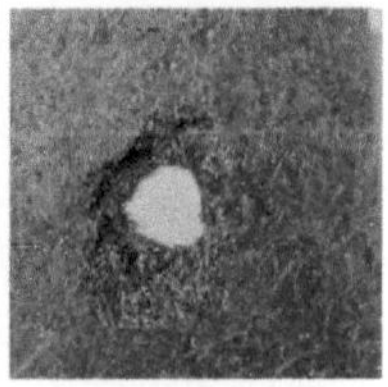

Table 4 Surface damage of CFBTS after subjected to impact test at different impact energy

CFBTS	Impact energy (J)		
	5	10	15
Front view			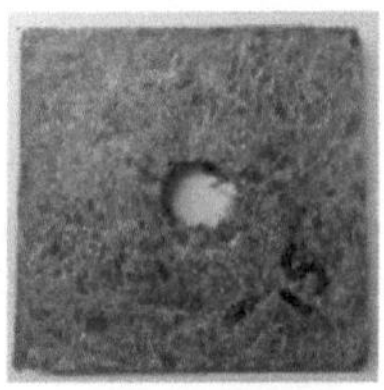
Back view			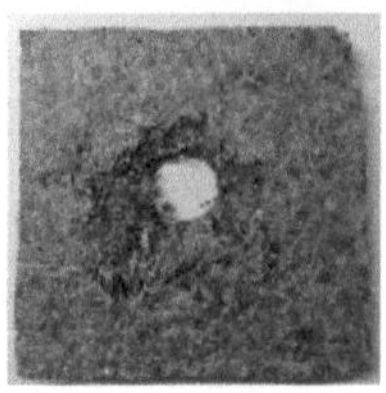

4 Conclusion

The drop weight impact behavior of CFB based natural binders namely CFBCS, CFBTS and CFBRF was successfully fabricated, characterized and compared. The CFBCS, which used corn starch, exhibited the highest peak load, energy absorbed and impact strength followed by CFBTS and CFBRF which used tapioca starch and rice

Table 5 Surface damage of CFBRF after subjected to impact test at different impact energy

CFBRF	Impact energy (J)		
	5	10	15
Front view			
Back view		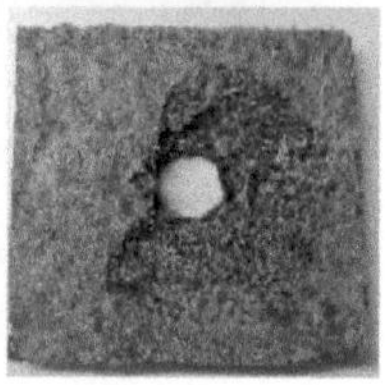	

flour, respectively. The CFBRF had poor adhesion bonding between rice flour and coconut fiber; thus, it reduced the ability to absorb more energy as the surface damage showed a large damage area compared to other CFB. The carbohydrate content within the natural binder is the main reason to influence the adhesion bonding of CFB. The impact properties of peak load, energy absorbed and impact strength increased with the increase of impact energy level for all types of CFB.

Acknowledgements The author acknowledges Universiti Teknologi MARA, Shah Alam, Selangor for funding under 600-RMC/MYRA 5/3/LESTARI (058/2020) and 600-RMC/DINAMIK-POSTDOC 5/3 (004/2020) grants.

References

1. Abbott AP, Conde JP, Davis SJ, Wise SJ (2012) Starch as a replacement for urea-formaldehyde in medium density fibreboard. Green Chem 14(11):3067–70
2. Aisyah HA et al (2013) Properties of medium density fibreboard (MDF) from kenaf (Hibiscus Cannabinus L.) core as function of refining conditions. Compos Part B Eng 44(1):592–96
3. Alotaibi JG, Alajmi AE, Yousif BF, Salih ND (2020) Effect of fibre content on compressive and flexural properties of coconut fibre reinforced epoxy composites. Am J Appl Sci 17(1):141–55
4. Garcia CM, San MR, Pablo SB, Pichelin F (2018) Resistance of fibreboards made of milled coconut husk and bonded with tannin against insects and fungi under philippine conditions. Int Wood Prod J 9(3):99–107
5. Kariuki SW, Wachira J, Kawira M, Murithi G (2019) Formaldehyde use and alternative biobased binders for particleboard formulation: a review. J Chem
6. Kim T (2019) Production planning to reduce production cost and process using medium-density fiberboard. Processes 7:529

7. Manek RV, Builders PF, Kolling WM, Emeje M, Kunle OO (2012) Physicochemical and binder properties of starch obtained from cyperus esculentus research article physicochemical and binder properties of starch obtained from cyperus esculentus. AAPS PharmSciTech 13:2
8. Ngadiman N et al (2018) Panel board from coconut fibre and pet bottle. In: E3S web of conferences, vol 34, pp 1–9
9. Antov P, Krišt'ák L, Réh R, Savov V, Papadopoulos AN, Çoban O (2021) Eco-friendly fiberboard panels from recycled fibers bonded with calcium lignosulfonate. Polymers 13:639
10. Radosavljevic L, Pecenka R (2008) Natural binders for fibreboard made of hemp. In: International conference on flax and other bast plants
11. Šafaric R, Zemljic LF, Novak M, Dugonik B, Bratina B, Bolka S, Gubeljak SSN (2020) Preparation and characterisation of waste poultry feathers composite fibreboards. Materials 13:4964
12. Sun J, Li L, Cheng H, Huang W (2017) Preparation, characterization and properties of an organic siloxane modified cassava starch-based wood adhesive. J Adhes
13. Tajuddin M, Ahmad Z (2016) A review of natural fibers and processing operations for the production of binderless boards. BioResourses 11(2):5600–5617
14. Wang ZTHJ, Luo L, Zhang J, Zhang HW, Chen HJ (2020) Mechanical properties of basalt fiber improved by starch phosphates sizing agent. Appl Surf Sci 521:146196
15. Takagi YRH, Nakagaito AN (2016) Tensile and flexural properties of polylactic acid-based hybrid green composites reinforced by kenaf, bamboo and coir fibers. Ind Crops Prod 94:562–573
16. Qiao Z, Gu J, Lv S, Cao J, Tan H, Zhang Y (2016) Preparation and properties of normal temperature cured starch-based wood adhesive. BioResourcesoresources 11(2):4839–49

Palm Byproducts Polymeric Composites

Processing and Alkali Treatment Impact Towards Oil Palm Frond Fibers Bulk Density and Wood-Plastic Composite Performance

Nor Yuziah Mohd Yunus, Nor Farhana Jasmi, and Wan Mohd Nazri Wan Abdul Rahman

Abstract The oil palm biomass is challenging in term of its mass utilization and its compatibility to various processing. The use of alkali treatment on wood fibers could modify the surface, thus making it more suitable for processing with polypropylene. This work analyses the impact of alkali treatment on the bulk density and its impact on the mechanical properties of wood-plastic composite (WPC). Fibers used were treated at 3 concentration level of sodium hydroxide: NaOH (1, 2, and 4%) and compared to control (no treatment). Two fiber loading 10 and 50 wt.% were blended representing high matrix and high fiber environment respectively. The materials were blended at 180 °C for 40 min and palletized. The test samples were pressed at 1000psi with temperature of 195 °C for 6 min. Test samples were prepared and tested in accordance to relevant ASTM procedure. Bulk density of fibers indicated an inversely proportional relation to concentration of alkali. Mechanical and physical properties showed better performance after alkali treatment and the impact was varied according to the loading factor of the composite. While tensile and flexural modulus of rupture plus elongation was at higher value for 10 wt.% fiber loading, the composite with 50 wt.% fibers exhibited higher modulus of elasticity and impact performance. The trend of 10 wt.% loading was either V shaped or inverted V shaped for mechanical properties indicating the impact at 2% modification being strongest. The 50 wt.% fiber loaded WPC have upward trend throughout the mechanical properties proportional to the alkali content. Capillary action of fiber in void created in higher loaded composite explained the lower absorption for 10 wt.% loaded WPC. Composites using treated fiber are a good option for future development as it could be further optimized using varying processing parameters such as temperature, pressure, time, loading factor, and coupling agents addition.

N. Y. M. Yunus (✉) · W. M. N. W. A. Rahman
Universiti Teknologi MARA Cawangan Pahang, 26400 Jengka, Pahang, Malaysia
e-mail: noryuziah@gmail.com

W. M. N. W. A. Rahman
e-mail: wmdnazri@uitm.edu.my

N. F. Jasmi
Faculty of Applied Sciences, Universiti Teknologi MARA, 40450 Shah Alam, Selangor, Malaysia

M. Jawaid et al. (eds.), *Proceedings of 2nd World Conference on Byproducts of Palms and Their Applications*, Springer Proceedings in Materials 19,
https://doi.org/10.1007/978-981-19-6195-3_6

Keywords Alkali treatment · Bulk density · Oil palm frond · Wood-plastic-composite

1 Introduction

Historical utilization of wood as building material centers on the material processing advantages of high physical and mechanical strength, low processing cost with aesthetically pleasing character [1]. Its major disadvantage of being prone to deterioration by environmental variations can be reduced by being married to plastic, where the blend would take the middle character associated to both materials [2]. Panel board made of 100% polypropylene (PP) material is costly due to its favored utilization compared to the other plastic matrices. The high cost could backfire as unfilled PP could make it less affordable to PP user. Filling polypropylene with natural filler is possibly an efficient way to decrease the utilization of PP material. Reduced consumption of PP extends the resource thus leading to lower overall raw prices. The composites have advantages such as good fatigue resistance, good chemical resistance, provide environmental protection, low cost, low carbon dioxide emission and stress cracking resistance, ease of machining and lower manufacturing energy.

Generation of solid biomass in Malaysia's palm oil industry hovers around 110 million dry ton of solid biomass by year 2020 [3] due to the 90% production of waste from oil palm to 10% production of oil ratio [4]. Five types of oil palm biomass, namely oil palm fronds (OPF), oil palm trunks (OPT), empty fruit bunches (EFB), mesocarp fibers and palm kernel shell could be generated. While some are recycled for nutrients (shredded and allowed to degrade) and traditional open burning is not a favorable future option [5]. Oil palm frond (OPF) made up 46–49% of the solid biomass generated. That translates to about 55 million dry ton of oil palm frond in 2020. The massive yield of OPF from the oil palm plantation is easily obtained during replanting and harvesting (twice a month), or pruning process. For each palm, average annual pruning contribution is approximately 82.5 kg of fronds [6]. The oil palm frond (OPF) is potential lignocellulosic filler in plastic matrix such as polypropylene (PP). When processed and made into plastic pellets which could be moulded to shapes, the material opens option for processing using standard plastic molding machine [7]. The combination of plastic with biomass that would otherwise be treated as waste makes it more inviting.

To form good interface between the two components, the hygroscopic lignocellulosics need to be compatibilized with the hydrophobic polymer matrix. Encapsulation properties of polymer matrix towards the lignocellulosic filler could be enhanced via filler surface chemical treatments, addition of coupling agent and radiation. For surface modification of the natural filler, utilization of sodium hydroxide (NaOH) solution is preferred as means to improve adhesion between filler and polypropylene matrix [8, 9]. This is based on its availability and associated cost. NaOH may degrade the cementing materials, oils, waxes and impurities from the filler surface thus exposing the crystalline structured cellulose [10]. NaOH reacts with the hydroxyl groups of hemicellulose to split the fiber bundles into filaments and results in better particle dispersion [8]. Lignocellulosic fibers such as hemp, banana leaves, jute and sisal have successfully been used as reinforcing materials in many plastic matrices. In this study, NaOH treatment was conducted to modify the lignocellulosic OPF filler surface in order to improve the adhesion between the filler and polymer matrix.

2 Material and Method

Oil palm fronds were obtained while harvesting the fresh fruit bunches of 12 years old oil palm tree. The OPF leaves were manually stripped, and the stalk portioned to top, middle and bottom to determine their physical and chemical properties. The stalks were chipped, flaked and ground to obtain dust of mesh 60 for chemical analysis. The dust was treated with 0, 1, 2 and 4%. For WPC the stalks were chipped, flaked, air dried, screened and fine <0.5 mm were sieved into three sizes: 75, 250, and 425 μm. These were then oven dried at 80 °C for 24 h prior to compounding process for composite preparation.

Alkaline treatment involved a procedure of immersing the oil palm frond particles in NaOH at 30 °C for 60 min in a rotary digester. The particles were washed to pH 7 with distilled water, followed by 24 h oven drying at 80 °C. The dried particles were sieved and kept in sealed container till compounding. Two loading 10 and 50 wt.% was prepared in a dispersion mixer with helical blade rotating at the speed of 10 rpm at temperature of 180 °C for 40 min. Test samples were cut from flat pressed composite prepared at 1000 psi using with press temperature of 195 °C for 60 s. The finished composite boards were conditioned and cut to relevant test specimen for flexural, tensile, impact, water absorption and thickness following ASTM requirement. The board cutting was performed in accordance to [11] for flexural properties of unreinforced and reinforced plastics, [12] for tensile properties of polymer matrix composite materials, [13] for Izod pendulum impact resistance, and [14] for water absorption. Once prepared, the specimens were conditioned at 21 °C ± 2 with relative humidity of 65% ± 5 for at least 24 h. All results were analyzed using Statistical Analysis System (SPSS) software.

3 Results and Discussion

Chemical Properties of Oil Palm Frond. Chemical analysis results on the OPT particles with NaOH treatments at concentrations of 0, 1, 2, and 4% are given in Table 1. The chemical compositions extracted from the experiment consist of cold water soluble, hot water soluble, alcohol toluene soluble, lignin, ash, holocellulose, and α-cellulose contents. The effect of 1%NaOH treatment has resulted in a declined trend for all the chemical components content except for the cellulosic alpha-cellulose component. All of the results obtained are comparable and in the

Table 1 Chemical properties of oil palm frond with and without treatment

Chemical compositions	NaOH concentrations (%)			
(%)	0	1	2	4
Cold-water solubility	14.67(0.60)	3.84(0.40)	3.68(0.25)	4.88(0.67)
Hot-water solubility	15.59(0.34)	3.71(0.37)	4.24(0.20)	5.38(0.73)
Alcohol-toluene solubility	5.25(0.45)	2.22(0.10)	1.91(0.11)	1.49(0.03)
Lignin content	23.65(0.42)	21.56(0.52)	21.44(0.61)	19.75(0.37)
Ash content	4.00(0.08)	2.36(0.05)	2.97(0.16)	3.34(0.02)
Holocellulose content	69.50(0.30)	69.32(0.28)	68.24(0.25)	65.00(0.61)
Alpha-cellulose contents	49.41(0.23)	50.15(1.01)	50.83(0.45)	54.59(0.17)

Note The values in parentheses are standard deviation

range of values described in literature by [15]. Cold water solubility decreased with increasing NaOH concentration from 0 to 2%. Both cold and hot water soluble extractives were the most affected by the alkali treatment with up to 70% reduction as compared to the untreated sample. However, when the concentration was brought up to 4%, the water soluble extractive contents were slightly increased. Similar trend of the slight increment at 4%NaOH concentration was seen in the ash content analysis.

The ANOVA done on chemical properties of oil palm frond indicated all the NaOH concentration having significant effect on all chemical properties at $p \leq 0.05$. Both hot water and cold water solubility extractive medium resulted in similar negative readings. The maximum value of cold water soluble was at 14.67% by untreated sample and followed by apparent decline values of treated samples. It also can be observed that, the alkali concentration has a significant effect on the hot water solubility. The hot water eliminates a great quantity of unwanted water soluble non-cellulosic materials from the compacted cellulose. It also removes a portion of the cell structure and extracts some inorganic extractives [16].

The alcohol toluene solubility is used to determine the extractive content that need chemical reagent for its extraction. Its means are reduced with the increasing of NaOH concentrations (Fig. 1). The extractive reduction was indeed necessary for the OPF filler in preparation for blending with polymer matrix. It was anticipated that compatibility between the OPF filler and PP in thermoplastic composite will be better. Figure 1 also clearly indicates decrease in percentage of lignin content with increases of NaOH concentration. When OPF particles were subjected to the surface modification, the lignin was depolymerized, giving rise to sugars and phenolic compounds that are soluble in water [17, 18]. As the concentrations of NaOH increase, more sodium (Na^+) and hydroxyl (OH^-) ions are available to react with the substances on the OPF particle causing greater depolymerized lignin to leach out [19].

Bulk Density of Oil Palm Frond. Table 2 shows the bulk density of OPF at different particle sizes and NaOH concentrations. Increasing the NaOH concentrations have lessened the bulkiness as observed for 425 μm particle size. Despite of a slight increase of bulk density by 75 and 250 μm particle size at 4%NaOH, the increment

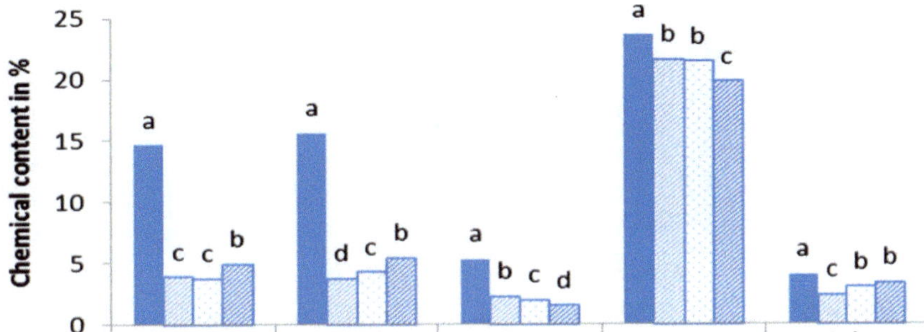

	CW	HW	AT	LC	Ash
0%NaOH	14.67	15.59	5.25	23.65	4
1%NaOH	3.84	3.71	2.22	21.56	2.36
2%NaOH	3.68	4.24	1.91	21.44	2.97
4%NaOH	4.88	5.38	1.49	19.75	3.34

Fig. 1 Effect of NaOH concentrations on non-cellulosic components of oil palm frond fibre. *Note* a, b, c, d Different letters on top of the bar chart represents significant difference among the four NaOH concentrations ($p \leq 0.05$). CW, cold water; HW, hot water; AT, alcohol toluene; LC, lignin content; Ash, ash content

of alkali concentrations have actually resulted in decreasing the bulk density of these two particle sizes too.

Effect of NaOH concentrations on bulk density of OPF is displayed in Fig. 2. At 4%NaOH, particle size of 75 μm was observed to have the largest bulk density followed by 250 and 425 μm particle size. Untreated particle gave higher bulk density as compared to all treated samples. The alkali treatment cause primary cell wall damage, making the inter-fibrillary region less dense and thus reduces the total weight of the filler per cubic meter [20]. NaOH treatment has reduced hemicellulose, lignin, waxes, oils and surface impurities from the OPF surface. Oil palm frond particle surface modification by NaOH solution has resulted in lowering the bulk density of the filler but without any further difference when the NaOH concentrations increase.

In term of processing performance, low bulk density may restrain the flow of material loading especially in large scale industrial consumption. This is because low bulk density may increase the transportation costs [21]. The feasible supply

Table 2 Summary of bulk density at different particle sizes and NaOH concentrations

Bulk density	NaOH concentrations (%)			
	0	1	2	4
425 μm	268.00(2.37)	148.65(2.75)	144.60(1.62)	142.00(1.60)
250 μm	277.73(4.50)	151.93(2.96)	141.50(2.00)	163.70(4.80)
75 μm	247.85(3.30)	150.58(2.05)	145.70(1.68)	175.28(3.65)

Note The values in parentheses are standard deviation

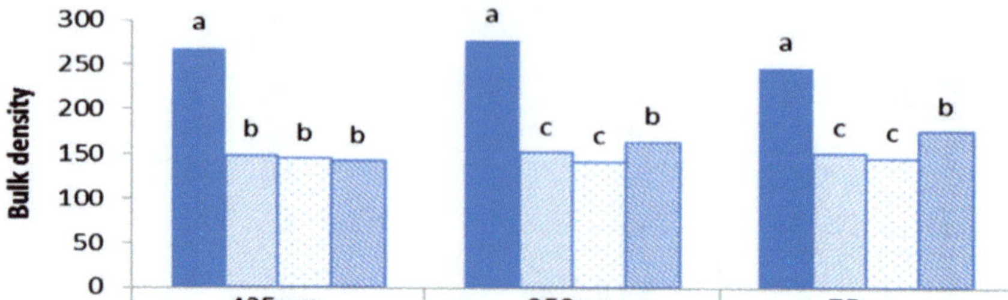

	425µm	250µm	75µm
0%NaOH	268	278	248
1%NaOH	149	152	151
2%NaOH	145	142	146
4%NaOH	142	164	175

Fibre size

Fig. 2 Effect of NaOH concentration on bulk density of different sizes of oil palm frond fibre. *Note* a, b, c different letters in group represent significant difference at $p \leq 0.05$

basis for the material having low bulk density may be only half of the required amount in order for processing and transporting works. This is due to the limit of the materials compaction. Moreover, high bulk density of natural particle may have higher tendency to be accumulated and formed bridging at some location while in the process of fabricating the composite. This accumulation problem was observed in a study that used the bulky date palm petiole's fiber during high fiber content of 40 wt.% [21]. The problem further may restrain the composite from achieving optimum strength.

ANOVA on bulk density of OPF at different particle size and NaOH concentration and the interaction of those two variables shows significant effects on bulk density at $p \leq 0.05$. For particle size with 2 ° of freedom, F-value was 28.345** and for NaOH concentration the F-value registered 44.09** (3° of freedom). The combination of size and NaOH treatment showed F-value of 70.017**. This indicates that combination of size and NaOH treatment need to be optimized for a particular targeted bulk density.

Effect of NaOH Treatment on OPF-PP Composite Properties. Mechanical and physical properties (mean and standard deviation) for 10 and 50 wt.% OPF in PP at different alkali concentrations are summarized in Table 3. In general, the composite exhibits apparent increment of both tensile strength and modulus properties but inconsistent pattern of elongation at break. Generally, it could be seen that the increment of NaOH concentration have inverse relation to water absorption.

ANNOVA of the properties (Table 4) shows the significant effects of the two independent variables upon all the composite properties. In the meantime, the effects of interaction between the independent variables varied from each other. The interaction between alkalization and filler loading effects gave insignificant difference at flexural modulus of elasticity, tensile modulus of elasticity and elongation at break.

Table 3 Summary of mechanical and physical properties in relation effects of filler loading and NaOH concentration

FL wt.%	A %	FMOR [Mpa]	FMOE [Mpa]	Impact (kJ)	TMOR [Mpa]	TMOE [Mpa]	Elong (%)	WA (%)
10	0	50.98 (0.37)	1888 (19)	4.05 (1.02)	16.17 (1.72)	1411 (76)	5.91 (0.83)	0.49 (0.11)
	2	41.99 (1.64)	2289 (67)	2.42 (1.02)	24.67 (0.31)	2520 (84)	5.99 (0.34)	0.37 (0.07)
	4	49.16 (2.33)	2265 (63)	4.63 (0.26)	22.55 (2.32)	2389 (51)	6.58 (0.81)	0.26 (0.04)
50	0	26.91 (0.57)	2566 (371)	5.52 (0.66)	8.26 (1.01)	1773 (76)	3.13 (0.05)	4.92 (0.46)
	2	28.35 (0.69)	3212 (271)	6.92 (0.10)	11.53 (0.22)	3199 (161)	3.25 (0.07)	3.99 (0.13)
	4	34.36 (1.26)	3529 (28)	7.14 (0.10)	14.51 (0.09)	3600 (168)	3.52 (0.21)	3.68 (0.56)

Note The values in parentheses are standard deviation. FL; filler loading, A%; NaOH concentration; FMOR—Flexural Modulus of Rupture; FMOE—Flexural Modulus of Elasticity; TMOR—Tensile Modulus of Rupture; TMOR—Tensile Modulus of Elasticity; Elong—elongation; WA—Water Absorption

Table 4 Summary of ANOVA of the filler loading and NaOH concentration effects on OPF-PP composite properties (F-values)

	df	FMOR	FMOE	TMOR	TMOE	Elong	Impact	WA
NaOH (A)	2	22.611**	12.895**	76.201**	78.539**	2.629**	21.165**	26.216**
Filler (F)	1	447.122**	987.417**	441.321**	429.570**	432.505**	221.385**	2087.431**
A × F	2	8.595**	1.448ns	10.356**	2.611ns	1.493ns	12.288**	9.763**

Note Df—degree of freedom; **—significantly difference at $p \leq 0.05$ level; ns—not significant at 0.05; FMOR—Flexural Modulus of Rupture; FMOE—Flexural Modulus of Elasticity; TMOR—Tensile Modulus of Rupture; TMOR—Tensile Modulus of Elasticity; Elong—elongation; WA—Water Absorption

Figure 3 shows the effect the flexural behavior of 10 and 50 wt.% OPF-PP composite under different alkali treatments 0, 2 and 4%. For 10 wt.% loading, the utilization of 2%NaOH concentration is insufficient to improve the flexural strength of the OPF-PP composite and even caused detriment to the composite strength. NaOH treatment should have improved the OPF particle dispersion as theoretically it increase the availability of hydroxyl groups on the OPF surface. However, the improvement of particle dispersion may be an independent factor for increasing the bonding between the natural filler and polymer matrix [22]. The availability of the hydroxyl groups of the low filler content of 10 wt.% is seen to be insufficient to entangle with the 90 wt.% polypropylene matrix content. The flexural strength is similar to 0% treatment when the alkali concentration increases to 4%. The 50 wt.% OPF-PP showed significantly improved Flexural Modulus of Rupture (FMOR) in proportional manner. Removal of extractive gave a more compatible surface for interaction with performance impact more prominent with used of higher fiber dosage. Flexural Modulus of Elasticity (FMOE) for both loading showed higher flexibility with treatment. These was comparable to extractives removal from fibers for both loading significant at $p \leq 0.05$. Further change from 2 to 4% aOH treatment does not have significant impact.

Comparing to the strength of composite at 10 wt.%, this reading gave us a small conclusion that the treatment involving NaOH solution on OPF-PP composite was more efficient at higher filler loading. Alkali treatment has produced a shrinkage and toughness effect on the natural filler that inbuilt a chemical strength between natural filler and polymer matrix [23].

Figure 4 exhibits the mean values of tensile properties of OPF-PP composites. The Tensile Modulus of Rupture (TMOR) for both loading showed increasing strength as the NaOH concentration increase. For lower fiber loading the degree of treatment showed not significant impact, while 50 wt.% composite has proportional relationship. For both instances approximately 38% increment of tensile strength was seen compared to the control sample. The trend of relationship for both loading in Tensile Modulus of Elasticity (TMOE) was similar with directly proportional increment in

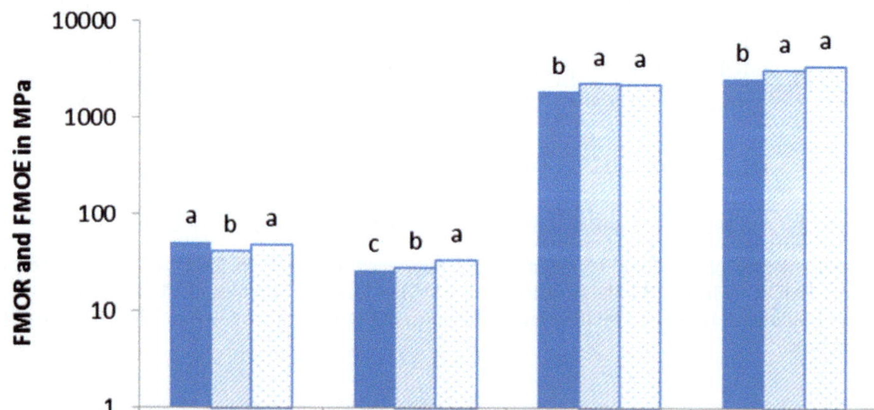

	10wt% FMOR	50wt% FMOR	10wt% FMOE	50wt% FMOE
0%NaOH	50.98	26.91	1888	2566
2%NaOH	41.99	28.35	2289	3212
4%NaOH	49.16	34.36	2265	3529

Fig. 3 Effects of alkalization on flexural properties of 10 and 50 wt.% filler loading. *Note* a, b, c Different letters in group represent significant difference at $p \leq 0.05$

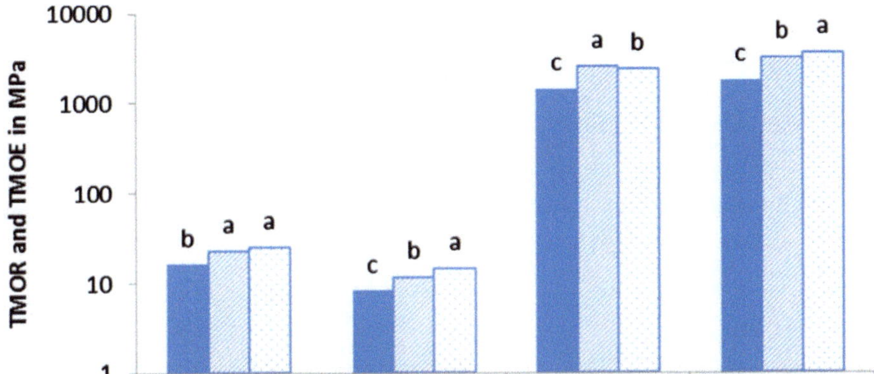

	10wt% TMOR	50wt% TMOR	10wt% TMOE	50wt% TMOE
■ 0%NaOH	16.17	8.26	1411	1773
▨ 2%NaOH	22.55	11.52	2520	3199
□ 4%NaOH	24.67	14.51	2389	3600

Fig. 4 Effects of alkalization on tensile strength of 10 and 50 wt.% filler loading. *Note* a, b, c different letters in group represent significant difference at $p \leq 0.05$

value as treatment percentage increase and the values were significant at $p \leq 0.05$. Similar trend of tensile strength improvement was observed on treated kenaf filled polypropylene with 3% and 6% NaOH concentrations [24]. This is attributed to the increment of uniformity of natural particles due to the removal of impurities on the filler surface by alkalization [25].

Bar chart (Fig. 5) on elongation at break behavior, showed the higher lengthening capability of 10 wt.% OPF-PP composites. The elongation at break increases in quantum with the increasing of NaOH concentration but with no significant difference. The response in elongation at break for 50 wt.% filler loading showed similar trending to 10 wt.% filler loading but at lower quantum. An apparent decrement of elongation at break was due to stiffness caused by the higher fiber content thus reducing the flow capability of the matrix. Meanwhile, higher elongation at break was actually a consequence of weak interfacial adhesion between the polymer matrix and natural filler as seen in the control sample [26].

Effect of alkalization treatment on impact strength of the OPF-PP composite shows a positive improvement (Fig. 6). The result shows an increment reading of impact strength with increases of alkali concentrations at 95% confident level intervals. The composite shows a modest decline result in impact strength when immersed in 2%NaOH but obtain better result when approaching the 4%NaOH. The effect of the clean surface caused by of NaOH treatment improved the filler and matrix adhesion. These changes caused increased of the impact strength [27]. Correlation analysis of filler loading and alkalization proves that filler significantly affect all properties and NaOH gives significant positive reading to TMOE and impact properties (Table 5).

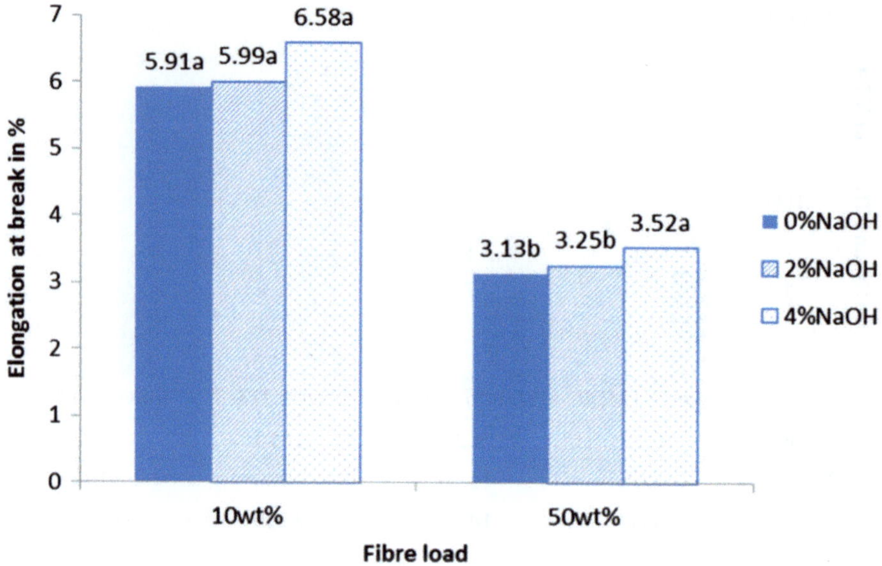

Fig. 5 Effects of alkalization on elongation at break at 10 and 50 wt.% filler loading. *Note* a, b different letters in group represent significant difference at $p \leq 0.05$

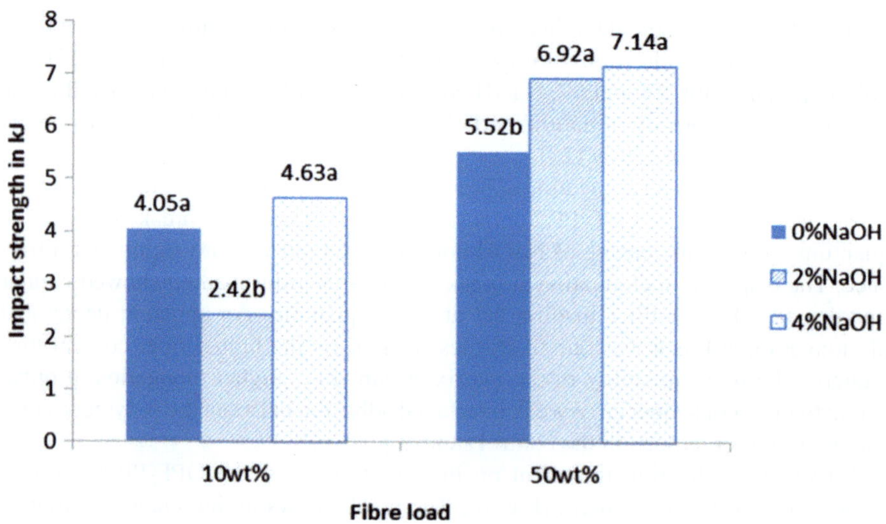

Fig. 6 Effects of alkalization on impact strength at 10 and 50 wt.% filler loading. *Note* a, b different letters in group represent significant difference at $p \leq 0.05$

Table 5 Correlation analysis of filler loading and alkalization

	FMOR	FMOE	TMOR	TMOE	Elong	Impact	WA
Filler loading	−0.742**	0.868**	−0.763**	0.616**	−0.928**	0.722**	0.879**
NaOH	0.098ns	0.117ns	0.406**	0.311**	0.140ns	0.267*	−0.203ns

Note ns—not significant; * significant difference at $p \leq 0.05$; ** highly significant difference at $p \leq 0.05$; FMOR—Flexural Modulus of Rupture; FMOE—Flexural Modulus of Elasticity; TMOR—Tensile Modulus of Rupture; TMOR—Tensile Modulus of Elasticity; Elong—elongation; WA—Water Absorption

Alkali treatment shows a significant impact on the water absorption behavior of the composite (Fig. 7). An apparent decrease of water absorption was observed when the alkali concentration increased. The pattern is the same for 10 and 50 wt.% loading. This may be attributed to the effect of alkali treatment that has resulted in increment of exposed alpha-cellulose in oil palm frond particle surface. The high crystallinity of the cellulose may prevent water from entering the filler surface hence leading to decrease of water absorption in the composite [28]. The 50 wt.% loaded composite has higher water absorption as the higher fiber content allows more capillary action, thus creating stronger bulking effect. The 2%NaOH seems to be the optimal concentration in gaining good water absorption behavior for the 50 wt.% OPF-PP composite. This is due to 4%NaOH having no significant difference compared to the 2%NaOH sample. For both instances the polymer fills the voids on the treated particle surface by replacing the hydroxyl groups with bonded chemical group of Na^+ and led to the hygroscopicity reduction of the oil palm frond particle. This reaction may contribute to better interfacial adhesion between the oil palm frond filler and polypropylene matrix. Better interfacial adhesion means that the polypropylene-natural filler network packed tighter and resulted in decreasing of water molecules diffusing through the free path between the natural filler and plastic matrix. Hence, the water uptake behavior may be reduced.

4 Conclusion

The NaOH treatment had shown impact on the chemical composition. A directly proportional reduction of content was shown for all instances except for the alpha-Cellulose. Resultant cleaned fiber has lower bulk density as removal of extractive and lignin create more voids on fiber. The OPF-PP composite made had shown positive results in the flexural, tensile, impact, elongation and water absorption properties. The 10 wt.% OPF-PP simulated effect of high matrix content and 50 wt.% OPF-PP exhibit high filler content environment.

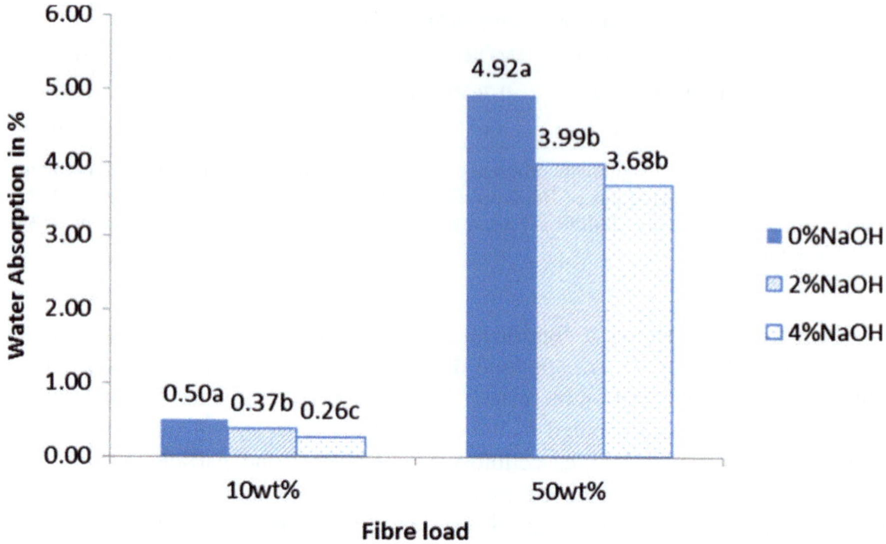

Fig. 7 Effects of alkalization on water absorption for 10 and 50 wt.% filler loading. *Note* a, b, c different letters in group represent significant difference at $p \leq 0.05$

References

1. Kuzman MK, Groselj P (2012) Wood as a construction material: comparison of different construction types for residential types for residential building using the analytic hierarchy process. Wood Res 57(4):59–600
2. Sumaila M, Ugheoke BI, Timon L (2006) A preliminary mechanical characterization of polyurethane filled with lignocellulosic material. Leonardo J Sci 9:159–166
3. Agensi Inovasi Malaysia (2013) National biomass strategy 2020: new wealth creation for Malaysia's biomass industry. http://www.nbs2020.gov.my/key-principles
4. Abdullah N, Sulaiman F (2013) The oil palm wastes in Malaysia, https://www.intechopen.com/chapters/44387
5. Singh P, Sulaiman O, Hashim R et al (2013) Using biomass residues from oil palm industry as a raw material for pulp and paper industry: potential benefits and threat to the environment. Environ Dev Sustain, 15367–383
6. Chan KW, Watson I, Kim LC (1981) Use of oil-palm waste material for increased production. In: Proceedings of the conference on soil science sand agricultural development in Malaysia, Malaysian Soil Science Society, Kuala Lumpur, pp 213–241
7. Teuber L, Osburg VS, Toporowski W et al (2016) Wood polymer composites and their contribution to cascading utilization. J Clean Prod 110:9–15
8. Cao Y, Shibata S, Fukumoto I (2006) Mechanical properties of biodegradable composites reinforced with bagasse fibre before and after alkali treatments. Compos Part A Appl Sci Manuf 37:423–429
9. Rokbi M, Osmani H, Imad A, Benseddiq N (2011) Effect of chemical treatment on flexure properties of natural fiber-reinforced polyester composite. Procedia Eng 10:2092–2097
10. Ichazo MN, Albano C, González J et al (2001) Polypropylene/wood flour composites: treatments and properties. Compos Struct 54(2–3):207–214
11. ASTM D790-03 (2003) Standard test methods for flexural properties of unreinforced and reinforced plastics and electrical insulating materials, ASTM International, West Conshohocken, PA

12. ASTM D3039/D3039M-00 (2006) Standard test method for tensile properties of polymer matrix composite materials, ASTM International, West Conshohocken, PA
13. ASTM D256-05a (2005) Standard test methods for determining the Izod pendulum impact resistance of plastics, ASTM International, West Conshohocken, PA
14. ASTM D570-98 (1998) Standard test method for water absorption of plastics, ASTM International, West Conshohocken, PA
15. Hashim R, Noor W, Wan A et al (2011) Characterization of raw materials and manufactured binderless particleboard from oil palm biomass. Mater Des 32(1):246–254
16. Shebani AN, van Reenen AJ, Meincken M (2008) The effect of wood extractives on the thermal stability of different wood species. Thermochim Acta 471:43–45
17. Erdocia X, Fernandez-Rodriguez J, Sequeiros A et al (2017) Lignin depolymerization process intensification by direct treatment of delignification black liquor. Chem Eng Trans 57:79–84
18. Fernández-Bolaños J, Felizón B, Heredia A et al (1999) Characterization of the lignin obtained by alkaline delignification and of the cellulose residue from steam-exploded olive stones. Biores Technol 68(2):121–132
19. Gu H (2009) Tensile behaviours of the coir fibre and related composites after NaOH treatment. Mater Des 30(9):3931–3934
20. Kim JK, Pal K (2010) Recent advances in the processing of wood-plastic composites. Springer, Berlin Heidelberg
21. Mahdavi S, Kermanian H, Varshoe A (2010) Comparison of mechanical properties of date palm fiber-polyethylene composite. BioResources 5(4):2391–2403
22. Beckermann GW, Pickering KL (2008) Engineering and evaluation of hemp fibre reinforced polypropylene composites: fibre treatment and matrix modification. Compos Part A Appl Sci Manuf 39:979–988
23. Mylsamy K, Rajendran I (2011) The mechanical properties, deformation and thermomechanical properties of alkali treated and untreated Agave continuous fibre reinforced epoxy composites. Mater Design 32(5):3076–3084
24. Meon MS, Othman MF, Husain H et al (2012) Improving tensile properties of kenaf fibers treated with sodium hydroxide. Eng Procedia 41:1587–1592
25. Edeerozey AMM, Akil H, Azhar AB, Ariffin MIZ (2007) Chemical modification of kenaf fibers. Mater Lett 61:2023–2029
26. John MJ, Francis B, Varughese KT, Thomas S (2008) Effect of chemical modification on properties of hybrid fiber biocomposites. Compos Part A Appl Sci Manuf 39:352–363
27. Punyamurthy R, Sampathkumar D, Bennehalli B et al (2014) Abaca fiber reinforced epoxy composites: evaluation of impact strength. Int J Sci Basic Appl Res (IJSBAR) 18(2):305–317
28. Liu R, Peng Y, Cao J, Chen Y (2014) Comparison on properties of lignocellulosic flour/polymer composites by using wood, cellulose, and lignin flours as fillers. Compos Sci Technol 103:1–7

Effect of Angle Ply on Tensile Strength of Unidirectional Glass/Epoxy and Arenga Pinnata/Epoxy Hybrid Composite Laminate

Jamaliah Md Said, Aidah Jumahat, and Jamaluddin Mahmud

Abstract Over the last few decades, researchers have aggressively investigated natural fibre reinforced polymer composites for replacing conventional synthetic polymeric materials in a variety of applications, including automotive, medical, agricultural, thermal management, and building insulation. In this study, modelling and simulation technique was used to predict the tensile properties of Unidirectional Glass/Epoxy and Arenga Pinnata/Epoxy Hybrid Composite Laminate. The effect of hybridization on the tensile strength of FRP composites was evaluated using ANSYS software based on maximum stress failure criteria. A symmetric finite element modelling of $[\pm\theta_G/\pm\theta_{AP}]_S$ hybrid FRP composites were analysed and compared to the properties of $[\theta_G/-\theta_G]_{2S}$ and $[\theta_{AP}/-\theta_{AP}]_{2S}$. Glass/Epoxy and hybrid with Arenga Pinnata/Epoxy natural fibre laminates subjected to uniaxial tension were simulated. The stress of the hybrid laminate composite of natural fibre from Arenga Pinnata yielded a moderate stress value, indicating that the composite was reliable. The maximum tensile stress of fibre laminates with the orientation of 0°, 15°, 30°, 45°, 60°, 75°, and 90° degrees were evaluated. Glass Fibre Reinforced Polymer (GFRP) fibres had the maximum uniaxial tension stress of 342.56 MPa at 0° and 78.38 MPa 30° ply orientation, while both fibres had the lowest uniaxial tension stress of 9.25 MPa and 7.20 MPa, at 90° ply orientation, respectively. Meanwhile, the hybrid of GFRP and Arenga Pinnata Fibre Reinforced Polymer (APFRP) had a

J. M. Said (✉)
College of Engineering, Universiti Teknologi MARA, Cawangan Johor, Kampus Pasir Gudang, Shah Alam, Malaysia
e-mail: jamaliahmdsaid@uitm.edu.my

A. Jumahat (✉) · J. Mahmud
School of Mechanical Engineering, College of Engineering, Universiti Teknologi MARA, 40450 Shah Alam, Selangor, Malaysia
e-mail: aidahjumahat@uitm.edu.my

A. Jumahat
Institute for Infrastructure Engineering and Sustainable Management (IIESM), Universiti Teknologi MARA (UiTM), Shah Alam, Selangor, Malaysia

M. Jawaid et al. (eds.), *Proceedings of 2nd World Conference on Byproducts of Palms and Their Applications*, Springer Proceedings in Materials 19,
https://doi.org/10.1007/978-981-19-6195-3_7

tensile property of 264.95 MPa at 0° and 4.53 MPa at 90°. This study contributes new knowledge in predicting the tensile strength value using modelling and simulation techniques.

Keywords Unidirectional Glass/Epoxy · Arenga Pinnata/Epoxy · Composite laminate · Hybridization · Uniaxial tension · ANSYS software

1 Introduction

Natural fibre hybrid polymer composites are increasing in popularity because of their recyclability, environmentalism, durability, and cost-effectiveness, among other characteristics. Creating fibre hybridization for different polymer matrixes may be accomplished in a variety of methods, such as mixing one synthetic fibre with one natural fibre or merging the fires of two separate natural fibres by using hybridization method for the optimum properties [1, 2]. The mixing of various natural fibre in a single polymer matrix resulted in improved mechanical characteristics when compared to single natural fibre reinforcement. Fibre hybridization for different polymer matrices can be done in a variety of ways, such as combining one synthetic fibre with one natural fibre or combining two different natural fibres. When compared to single natural fibre reinforcement, the combination of different natural fibres in a single polymer matrix improves the mechanical properties [3].

The mechanical properties of natural fibre are lower than the existing synthetic fibre in polymer composite material. Arenga Pinnata has properties compared to glass fibre. Therefore, the hybrid composite was made to improve the properties of the natural fibre with different arrangements in the direction of the fibre. Using polymer composites as a protection and recovery approach for fracture surfaces and pipe systems, polymer composites are becoming increasingly popular because of their inherent benefits over conventional fastener techniques. In comparison to other joint types, adhesive joints distribute stresses more uniformly, resulting in lower stress concentrations and higher strength to weight ratio in structures. Some researchers discovered that natural fibres, such as those from palm trees, may act as an adhesive [4–7]. When building laminated composite structures, the first ply failure load is an essential variable to consider. The development of theoretical and finite element analytic techniques in determining the first ply failure of laminated composites has been completed. According to Reddy et al., benchmark research was carried out to determine the first ply failure load of laminated composites using the finite element approach [8].

In this study, the properties of various Arenga Pinnata fibre layup sequences and hybrid design of the natural fibre reinforced polymer composites are predicted using modelling and simulation techniques. Ansys software is used to predict the tensile properties of Unidirectional Glass/Epoxy and Arenga Pinnata/Epoxy Hybrid

Composite Laminates. The effect of hybridization on tensile strength of FRP composites are evaluated using maximum stress failure criteria. A symmetric finite element modelling of $[\pm\theta_G/\pm\theta_{AP}]_S$ hybrid FRP composites are analysed and compared to the properties of $[\theta_G/-\theta_G]_{2S}$ and $[\theta_{AP}/-\theta_{AP}]_{2S}$. Glass/Epoxy and hybrid with Arenga Pinnata/Epoxy natural fibre laminates subjected to uniaxial tension are simulated.

2 Methods

2.1 Numerical Validation

To demonstrate that the original Finite Element (FE) modelling and implementation are acceptable, numerical validation is a critical early step. The use of FE software is critical, and the Ansys-based results are validated to ensure they correspond to the real results. The application of finite element analysis is carried out to relieve the time-consuming difficulty of establishing a strategic separation from time-consuming testing. Statistical verification is implemented for composite laminate plates with dimensions of 279 mm × 279 mm × 2.16 mm and a ply thickness of 5.5555 × 10^{-6} m. The plate is meshed using the SHELL281 element type, which has eight nodes. The current FE results from ANSYS have been validated against the precise solution. When subjected to transverse stresses, a composite laminar structure with a variety of maximum distortion caused by laminating patterns is estimated. The acquired research data is obtained by Soni [9]. The findings are satisfactory because the error is less than 2%.

2.2 Convergence Analysis

The determination of mesh size used for the modelling and simulation of unidirectional GFRP under tensile and compression loads is very crucial to ensure the accuracy of the results, hence improve the reliability of it. This is because, the greater the number of mesh size, the higher the accuracy of the value obtained. However, the greater the number of mesh size, the more amount of time is used for data computation. Besides, there is a limit size for meshing in ANSYS Mechanical APDL. For this convergence analysis, 2 × 2, 4 × 4, 8 × 8, 16 × 16, 32 × 32, 32 × 32, 64 × 64, 128 × 128, and 256 × 256 mesh sizes were tested on unidirectional glass fibre reinforced polymer (GFRP) for three angles of fibre which were 0°, 50°, and 90° under tensile load. The values of deformation remained the same even when the mesh size was increased. Therefore, mesh size of 8 × 8 was selected to be used for better accuracy of results and faster computation of data.

2.3 Modelling and Failure Analysis of Composite Laminates

The current work employs finite element programming that is publicly accessible (ANSYS 2020 R2) to receive and alter finite element modelling and simulation in exploring the displacement while First Ply Failure (FPF) load of composite laminates are predicted using Maximum Stress Criteria. This software is used to model the Unidirectional Glass/Epoxy and Arenga Pinnata/Epoxy Hybrid Composite Laminate due to different angle of fiber orientations under uniaxial tension load. The materials properties of Glass/Nanosilica Epoxy and Arenga Pinnata/Nanosilica Epoxy are given in Table 1 and Table 2, respectively. Pre-processing was a step done before the load was applied. First, the process started by modelling the specimen using Preprocessor. Under the Preprocessor, the area was defined using 2 corners area of rectangle. Second, the element type was selected by selecting Shell 8 node 281 which was used for thin layers of specimen. Third, material model behaviour and properties were defined using structural of elastic linear "Orthotropic".

After that, the laminates of specimen were defined using Shell lay-up depending on fibre orientation. In this study, the characteristics of APFRP, GFRP and hybrid of both fibre were applied with anti-symmetric stacking layup of $[\pm\theta_G/\pm\theta_{AP}]_{2S}$ hybrid FRP composites which were investigated and compared to those of $[\theta_G/-\theta_G]_{2S}$ and $[\theta_{AP}/-\theta_{AP}/\theta_{AP}/-\theta_{AP}]_S$ composites that were modelled in a symmetric approach. The ply thickness of glass fibre (GFRP) and Arenga Pinnata (APFRP) fibre used in this investigation was 0.25 mm and 0.425 mm, respectively. According to ASTM requirements, the size of the sample for each test is represented in Fig. 1a for the uniaxial tensile test, which is performed in line with ASTM standards. The schematic diagram and actual size of test sample with applied load are presented in Fig. 1b.

Table 1 Material properties for Glass/Epoxy [11]

Elastic parameter		Strength data	
E1	6.47 GPa	XT	342.56 MPa
E2 = E3	2.16 GPa	XC	77.29 MPa
v12	0.25	YT	9.249 MPa
G12 = G23 = G13	2.59 GPa	YC	10.35 MPa
–	–	S	41 MPa

Table 2 Material properties for Arenga Pinnata/Epoxy [12]

Elastic parameter		Strength data	
E1	4.142 GPa	XT	72.032 MPa
E2 = E3	0.414 GPa	XC	105.209 MPa
v12	0.25	YT	7.2 MPa
G12 = G23 = G13	2.59 GPa	YC	10.53 MPa
–	–	S	41.56 MPa

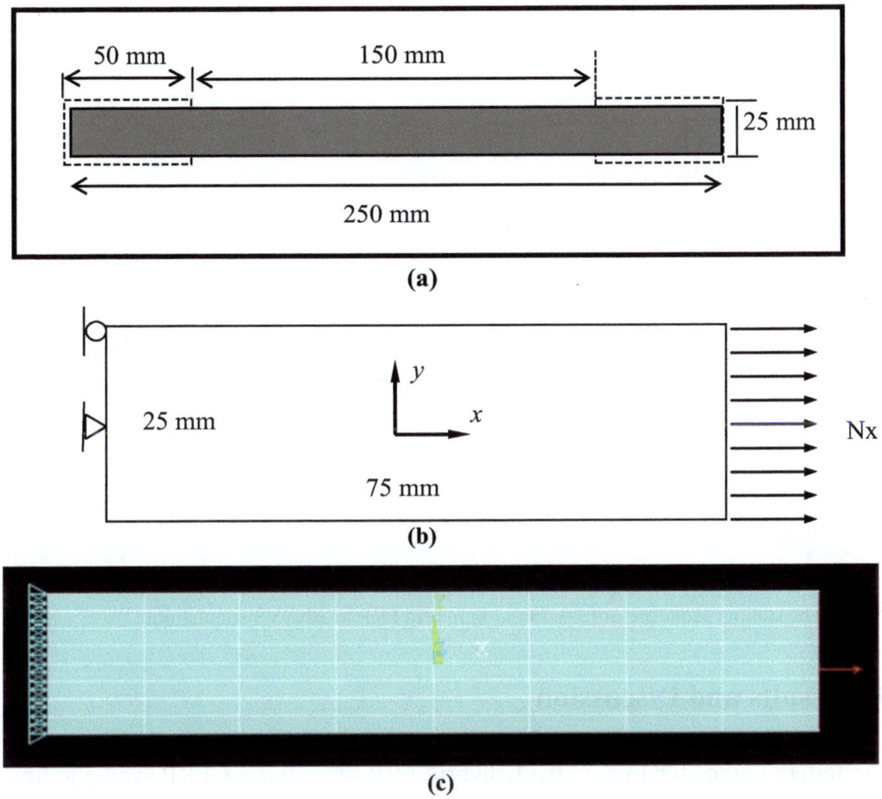

Fig. 1 **a** ASTM D3410: Specimen size for compression test [10] **b** Uniaxial tension sample with x-direction **c** Meshing, boundary conditions and applied load in ANSYS

Figure 1c represents the meshing and boundary conditions in ANSYS analysis with load applied.

For processing, displacements of structural at fixed points can be set equal to zero in Define Loads. For tensile, only one fixed point on one end of the specimen was set to zero. Next, the pressure applied on the specimen was set using Structural On-lines Pressure. Analysis was ready to start once the loads placement was finished using Solve Solution. Lastly, Post-processing was the result after analysis process. The behaviour can be observed through graphical and animation of Contour plot. Deformation and stress on all three x, y, and z can also be identified but only x direction displacement was measured for this study. Failure Criteria using Maximum Stress Theory was also obtained through this process. As part of this research, a mix of natural and synthetic fibre was examined, and the resulting layups are depicted in Fig. 2. With varied angle orientations, synthetic fibre was staked into the outer layer.

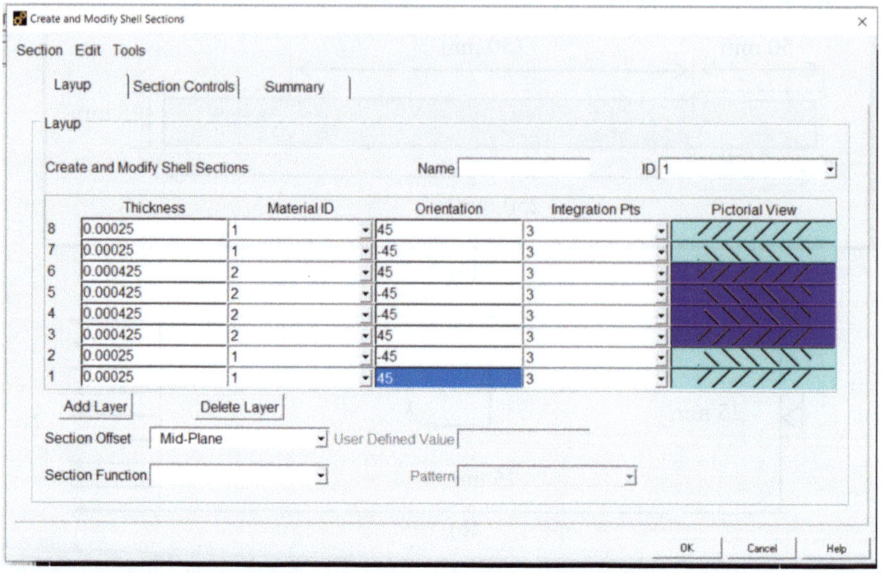

Fig. 2 The staking sequence of $[\pm\theta_G/\pm\theta_{AP}]_S$ hybrid FRP in ANSYS simulations

3 Results and Discussion

The failure curves for $[\pm\theta_G/\pm\theta_{AP}]_S$ hybrid FRP $[\theta_G/-\theta_G]_{2S}$ GFRP synthetic fibre and $[\theta_{AP}/-\theta_{AP}]_{2S}$ APFRP natural fibre composite laminate as predicted by Maximum Stress Theory are presented in Fig. 3. The failure patterns for both Glass Fibre Reinforced Polymer (GFRP) and Hybrid Fibre Reinforced Polymer (HFR) plates were identified, and in general, the trends of failure curvatures showed a gradual decrease, which means that the plates got weaker when the fibres were arranged away from the direction of the load. The failure curve for the natural fibre APFRP laminates, on the other hand, did not reveal any differences in terms of value of maximum stress for this layup technique. There was a gradual increase in the tensile strength of the natural hybrid [13]. The picture illustrates the differences between the synthetic and natural fibre conditions, as well as how the characteristics of the laminates composite improved when the two types of fibres were combined in a hybrid configuration. GFRP had the highest maximum value of stress in the 0° angle (342.56 MPa), while APFRP had the highest maximum value of stress in the 30° angle (78.38 MPa). A hybrid of synthetic fibre GFRP and natural fibre APFRP had the highest maximum value of stress (264.93 MPa) in the fibre direction 0° angle fibre orientation like GFRP fibre. When the direction of fibre orientation was at 90°, the minimal stress value for every type of fibre reinforced polymer composite was obtained, including 7.20 MPa (APFRP), 9.25 MPa (GFRP), and 4.53 MPa Hybrid FRP. An epoxy containing nanosilica was introduced into hybrid laminate composite materials that had been created by incorporating glass fibres into natural fibre. In this

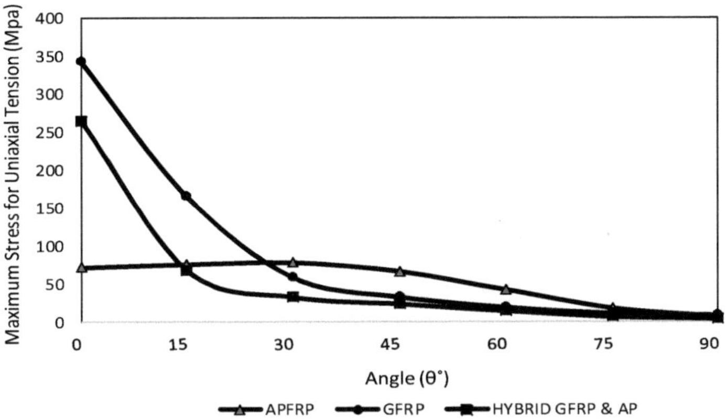

Fig. 3 Maximum stress of APFRP, GFRP and Hybrid (GFRP and APFRP) composites with different angle of fibre direction loaded in uniaxial tension

case, Arenga Pinnata fibres were treated with silane on their surfaces to improve the incorporation between natural fibre and epoxy. The staking sequence for this hybrid gave a moderate result between the Arenga Pinnata natural fibre and synthetic glass fibre. According to the findings of previous researchers, the staking of synthetic fibre at the outer layer resulted in improved strength outcomes [14, 15].

The results for the anti-symmetric dimensions plate laminate illustrate the deformation behaviour in x-directions on applied load during failure, which was consistent with the predictions of the Maximum Stress Theory. The displacement in the x-direction indicated that the maximum stress angles at 0° and 90° for GFRP synthetic fibre were 7.944 mm and 0.642 mm, respectively, and the maximum stress angles at 0° and 90° for Hybrid (GFRP and APFRP) fibre were 7.943 mm and 0.637 mm. As shown in Fig. 4, ARFRP natural fibre has the highest displacement at 30° fibre orientation with 2.992 mm while the lowest displacement occurred at 90° fibre orientation with 2.635 mm for the plate material. The difference was not significant since natural fibre was less elastic compared to synthetic fibre. However, by utilising glass and Arenga Pinnata as a hybrid between synthetic and natural fibres at a unidirectional angle for this consequence, it is possible to tailor the solution to the problem (see Fig. 4).

At failure load conditions of 0° and 90° fibre ply orientation, Fig. 5 illustrates the nodal solution on the maximum stress distribution with maximum applied load in the fibre ply for (a) Arenga Pinnata (APFRP) composite, (b) Glass Fibre (GFRP) synthetic fibre, and (c) Hybrid Fibre (GFRP and APFRP). As the applied load increased, the maximum stress continued to increase as well, resulting in a rise of displacement value. When examining the modelling results, the red colour represented the largest deflection that occurred at the right end of the sample since it was fixed on the left side. The parameters were like those used in experimental techniques, with the exception that the force was delivered perpendicular to the area of

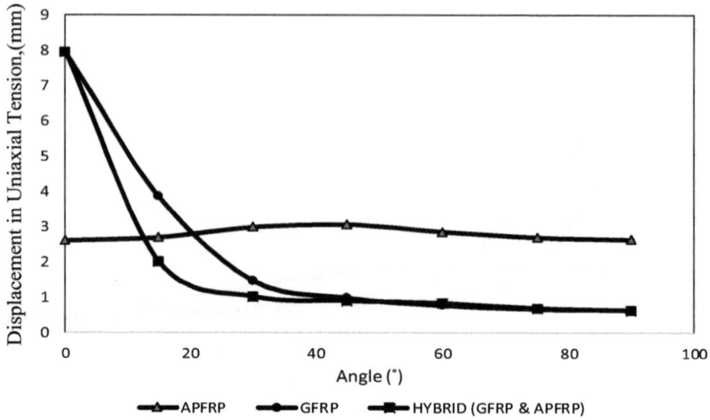

Fig. 4 Displacement in x directions of APFRP, GFRP and Hybrid (GFRP and APFRP) composites with different angle of fibre direction loaded in uniaxial tension

the sample. A sample's displacement value can be determined when it has extended in length relative to its initial length.

4 Conclusion

Hybridization of natural fibre reinforced composites such as Arenga Pinnata/Nanosilica epoxy, and synthetic fibre reinforced composites such as Glass/Nanosilica epoxy have their own specific set of properties which are hardness, fragility, and brittleness that may be attained by a combination of these two types of fibres. All these traits can lead to a pattern of consistent behaviour as well as failure. It has been decided to use a combination of synthetic and natural fibres to modify the brittleness of epoxy properties. To solve these concerns, Glass Fibre Reinforced Polymer (GFRP) and Arenga Pinnata Fibre Reinforced Polymer (APFRP) were laminated together in this research. Staking layups of $[\pm\theta_G/\pm\theta_{AP}]_S$ were used for hybrid FRP in comparison to $[\theta_G/-\theta_G]_{2S}$ APFRP and $[\theta_{AP}/-\theta_{AP}]_{2S}$ GFRP fibres. A GFRP fibre had the maximum direction uniaxial tension stress with 342.56 MPa when measured at 0° and 78.38 MPa at 30°, while both fibres had the lowest single-direction uniaxial tension stress when measured at 90° (9.25 MPa and 7.20 MPa respectively). Meanwhile, the hybrid between GFRP and APFRP had 264.95 MPa and 4.53 MPa at 0° and 90° respectively. Furthermore, when comparing natural fibre to synthetic fibre, hybrid techniques assist in raising the maximum stress that can be tolerated. Using finite element analysis, the maximum stress and load in a composite may be predicted for different angle fibre orientations without performing experimental results.

Nodal solution, θ = 0°	θ = 90°
(a) APFRP 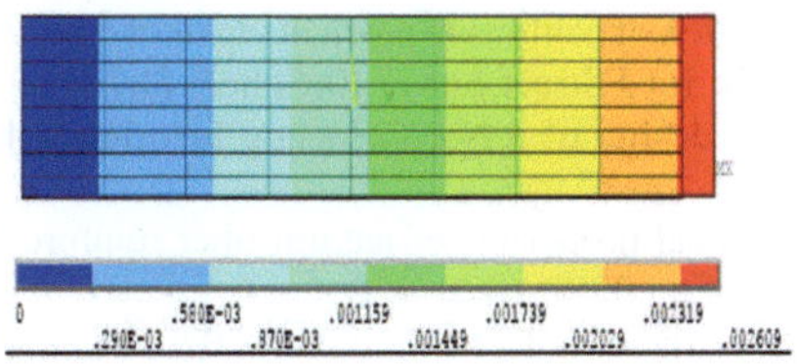 Failure load = 6122.65 N Maximum deflection = 2.609 mm	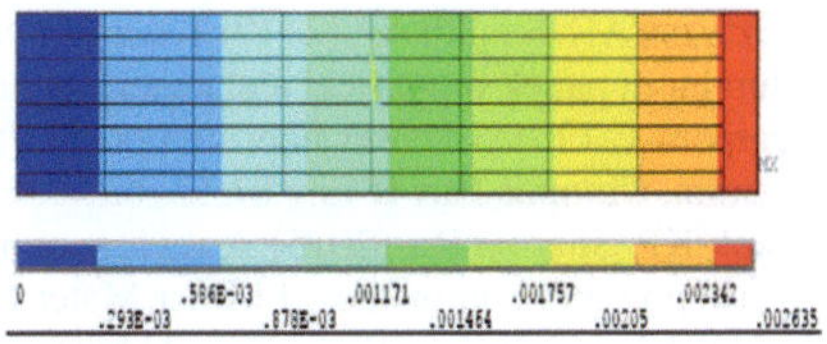 Failure load = 612.25 N Maximum deflection = 2.635mm
(b) GFRP 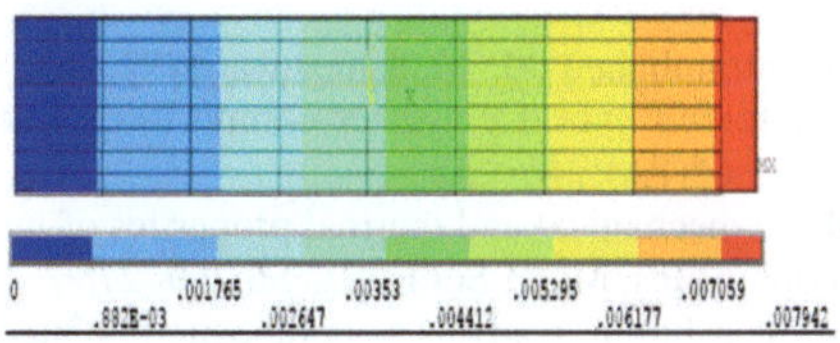 Failure load = 17127.98 N Maximum deflection =7.944 mm	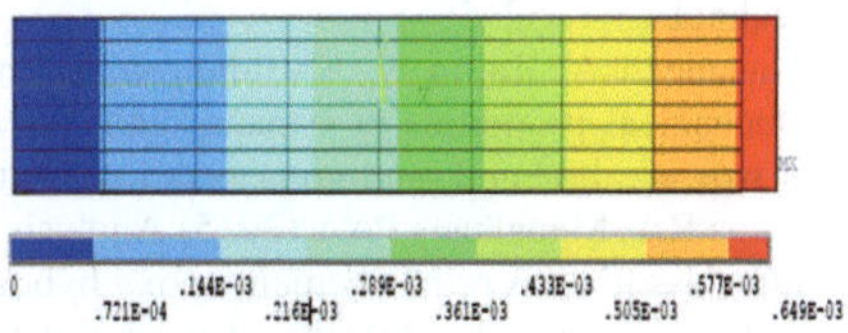 Failure load = 462.50 N Maximum deflection = 0.642 mm
(c) Hybrid (GFRP and APFRP) 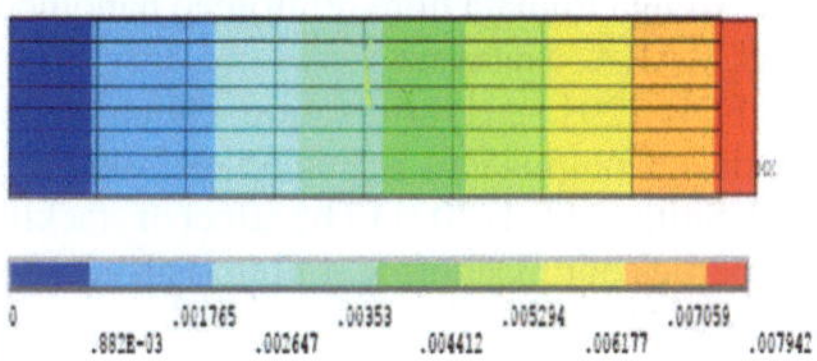 Failure load = 17883.97 N Maximum deflection = 7.943 mm	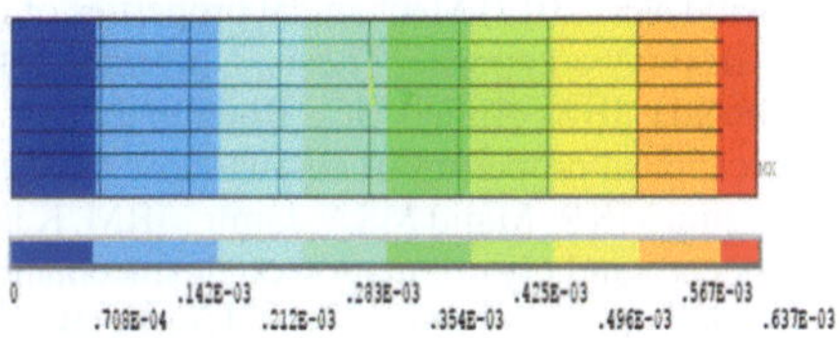 Failure load = 305.46 N Maximum deflection = 0.637 mm

Fig. 5 Deformation contour at failure load condition of 0° and 90° of **a** APFRP natural fibre **b** GFRP synthetic fibre **c** Hybrid (GFRP and APFRP) fibre

Acknowledgements The authors gratefully appreciate the financial support from Universiti Teknologi MARA (UiTM) and the Ministry of Higher Education, Malaysia. The research was performed at the School of Mechanical Engineering UiTM Malaysia under the support of LESTARI Grant No: 600-RMC/MyRA 5/3/LESTARI (058/2020).

References

1. Makwana AH, Shaikh AA (2019) Towards hybridization of composite patch in repair of cracked Aluminum panel: numerical and experimental study. Int J Struct Integr 10(6):868–887
2. Scarponi C (2015) Hemp fiber composites for the design of a Naca cowling for ultra-light aviation. Compos Part B Eng 81:53–63
3. Ashik KP, Sharma RS (2015) A review on mechanical properties of natural fiber reinforced hybrid polymer composites. J Miner Mater Charact Eng 03(05):420–426
4. Delzendehrooy F, Ayatollahi MR, Akhavan-Safar A, da Silva LFM (February 2020) Strength improvement of adhesively bonded single lap joints with date palm fibers: effect of type, size, treatment method and density of fibers. Compos Part B Eng 188:107874
5. Yahaya R, Sapuan SM, Jawaid M, Leman Z, Zainudin ES (2016) Effect of fibre orientations on the mechanical properties of kenaf–aramid hybrid composites for spall-liner application. Def Technol 12(1):52–58
6. Alavudeen A, Rajini N, Karthikeyan S, Thiruchitrambalam M, Venkateshwaren N (2015) Mechanical properties of banana/kenaf fiber-reinforced hybrid polyester composites: effect of woven fabric and random orientation. Mater Des 66(PA):246–257
7. Braga RA, Magalhaes PAA (2015) Analysis of the mechanical and thermal properties of jute and glass fiber as reinforcement epoxy hybrid composites. Mater Sci Eng C 56:269–273
8. Reddy YSN, Reddy JN (1992) Linear and non-linear failure analysis of composite laminates with transverse shear. Compos Sci Technol 44(3):227–255
9. Soni S (1983) A new look at commonly used failure theories in composite laminates
10. ASTM International, "Astm D3410-03," vol 08, pp 1–16
11. Hashim URBH (July 2020) Mechanical properties of hybrid graphene filled basalt fibre reinforced. Universiti Teknologi MARA, Thesis
12. Khalid MF (2017) Mechanical properties of treated Arenga Pinnata fibre reinforced nanomodified epoxy composite, Thesis, Universiti Teknologi MARA, Malaysia
13. Almeida JHS, Amico SC, Botelho EC, Amado FDR (2013) Hybridization effect on the mechanical properties of curaua/glass fiber composites. Compos Part B Eng 55:492–497
14. Hashim MKR, Majid MSA, Jamir MRM, Kasim FH, Sultan MTH (2021) The effect of stacking sequence and ply orientation on the mechanical properties of pineapple leaf fibre (Palf)/carbon hybrid laminate composites. Polymers (Basel) 13(3):1–24
15. de Queiroz HFM, Banea MD, Cavalcanti MD (2021) Adhesively bonded joints of jute, glass and hybrid jute/glass fibre-reinforced polymer composites for automotive industry. Appl Adhes Sci 9(1)

Bending Stress and Deformation Analysis of Nanosilica Filled Arenga Pinnata/Epoxy and Glass/Epoxy Polymer Composites

Ilya Izyan Shahrul Azhar, Aidah Jumahat, and Jamaliah Md Said

Abstract Bending in applied mechanics characterises the action of a slender structural structure subjected to an external load applied perpendicular to the element's longitudinal axis. Bending stress deformation behaviour and flexural properties of fibre reinforced polymer (FRP) composite materials are important in designing structures and mechanical components. The bending or flexural properties of FRP composite materials depend on the type of fibre and the fibre sequence architecture. In this study, bending stress and deformation analysis were determined using modelling and simulation techniques. The dimension of the specimen was set up based on ASTM standard D7264. The effect of fibre types, i.e., *Arenga Pinnata* and Glass, on flexural stress–strain behaviour, the effect of layer sequence on maximum bending stress and deformation properties, and the effect of 25 wt.% nanosilica inclusion in the epoxy were simulated and calculated. ANSYS Software was used to simulate the symmetric $[\theta/-\theta/0/0]_s$ laminate sequence. The effects of nanosilica and fibre orientation of 0°, 15°, 30°, 45°, 60°, 75°, and 90° on flexural behaviour were investigated. From this study, it was found that the 45° unidirectional laminate exhibited the highest flexural strength. The maximum bending stress of 25APFRP was 90.7 MPa with a maximum deflection of 4.447 mm. The presence of 5 wt.% nanosilica improved the bending properties of *Arenga Pinnata* with a maximum bending stress of 94.1 MPa and 4.486 mm maximum deflection. It can be concluded that FRP composites made of *Arenga Pinnata* FRP composites have high flexural properties when compared to conventional Glass FRP composites.

I. I. S. Azhar (✉) · J. M. Said
School of Mechanical Engineering, College of Engineering, Universiti Teknologi MARA, Cawangan Johor, Kampus Pasir Gudang, Shah Alam, Malaysia
e-mail: izyan0363@uitm.edu.my

A. Jumahat (✉)
School of Mechanical Engineering, College of Engineering, Universiti Teknologi MARA, 40450 Shah Alam, Selangor, Malaysia
e-mail: aidahjumahat@uitm.edu.my

Institute for Infrastructure Engineering and Sustainable Management (IIESM), Universiti Teknologi MARA (UiTM), Shah Alam, Selangor, Malaysia

M. Jawaid et al. (eds.), *Proceedings of 2nd World Conference on Byproducts of Palms and Their Applications*, Springer Proceedings in Materials 19,
https://doi.org/10.1007/978-981-19-6195-3_8

Keywords Bending · Flexural · Arenga Pinnata · Polymer composite · ANSYS

1 Introduction

Arenga Pinnata is an economically significant feather palm native to tropical Asia, from eastern India east to Malaysia, Indonesia, and the Philippines in the east. *Arenga Pinnata* has several common names which include sugar palm or black sugar palm. The main advantage of *Arenga Pinnata* fibres is that it has high durability and resistivity to seawater [1]. In fibre-reinforced polymer composite (FRPC), the fibre is a load-bearing member along its longitudinal axis, while the matrix distributes stress across the fibre reinforcements. Natural fibre has recently attracted the attention of scientists and technologists due to its outstanding qualities. Lightweight, low cost, environmentally friendly, high flexibility, renewability, biodegradability, high specific strength, good thermal properties, high durability, reduced tool wear, no skin irritation, and quick processing are all characteristics of natural fibres [2].

The advancement of materials has accelerated the creation of new materials. Composite material has developed into a major substance due to the benefits associated with the combination of more than two materials with distinct physical and chemical properties. The applications of composite materials include polymer as a matrix and reinforced with other materials in different shapes and sizes [3]. Polymer matrix composites have been studied and considered in many other applications with special properties of the composites [4, 5].

The matrix acts as a protective barrier from external and environmental damage and distributes the load among the fibres. The fibres, in turn, contribute strength and stiffness to the matrix, therefore strengthening it and making it more resistant to cracks and fractures [6]. The composite materials are developed to replace the conventional material with the reduced weight of the materials. One of the techniques to create composite materials with superior characteristics is by using a laminating process. The trend of laminating the fibre using a natural fibre as the reinforcement material has become a popular method in engineering applications [7]. The technique creates a composite laminate with the characteristics required for a particular purpose. This adaptability is precisely why composites are desirable for specific applications [8]. In addition to having high mechanical and chemical characteristics and corrosion resistance, epoxy resins also have outstanding thermal and dimensional stability and are widely utilized in lamination, adhesives, coatings, and advanced composite surfaces, among other applications.

Typically, a material's deformation criteria are determined experimentally using bending tests. However, nowadays, simulation and modelling techniques are alternative methods to predict various laminate composite designs, layup sequences, geometry, hybrid materials, and many others. Simulation technique has become one of the ideal solutions in determining the deformation and bending stress analysis of fibre epoxy composite laminates. By determining the angle of the lamination scheme, it is possible to estimate the deformation and bending stress analysis of the materials.

Therefore, this study aims to determine the bending stress and deformation analysis of fibre composite laminates into various types of composite laminates in multiple setups of layup sequences towards flexural and deformation profile via simulation analysis. The project revolves around synthetic and natural fibres. ANSYS software is utilized to simulate the bending fibre composite laminate; thus, gathering data on flexural behaviour and deformation of fibre composite laminates through the Finite Element Analysis (FEA) method. The presence of flexural behaviour upon composite laminates is also predicted in this study and compared to previous the work done by other researchers.

2 Materials and Method

The ANSYS software is a general purpose in the method of Finite Element Analysis (FEA) software which involves the numerical method of constructing a complex model into a smaller element size. The Finite Element Model (FEM) method using ANSYS for bending was modelled with specimen dimensions of 12.7 mm width, 3.4 mm thickness, and 100 mm length. In this study, symmetric angle-ply sequences of $[0/0]_{2s}$, $[\pm 15]_{2s}$, $[\pm 30]_{2s}$, $[\pm 45]_{2s}$, $[\pm 60]_{2s}$, $[\pm 75]_{2s}$, and $[90/90]_{2s}$ were chosen to be implemented across all types of fibre reinforced composite laminate sequences. The modelling and simulation of fibre composite laminates must be determined carefully to assure the accuracy of the results and hence increase their reliability. The larger the number of mesh sizes, the more accurate the result produced. However, the larger the number of mesh sizes, the longer it takes to compute data. The mesh elements and boundary conditions used for this study are illustrated in Fig. 1. The boundary conditions are applied at both ends while the load is applied in the middle line.

The first phase in the simulation is the pre-processor. The process of simulation and modelling was built by defining the material properties, modelling dimension, angle ply of the layer, and the failure criteria. The values of the ultimate tensile

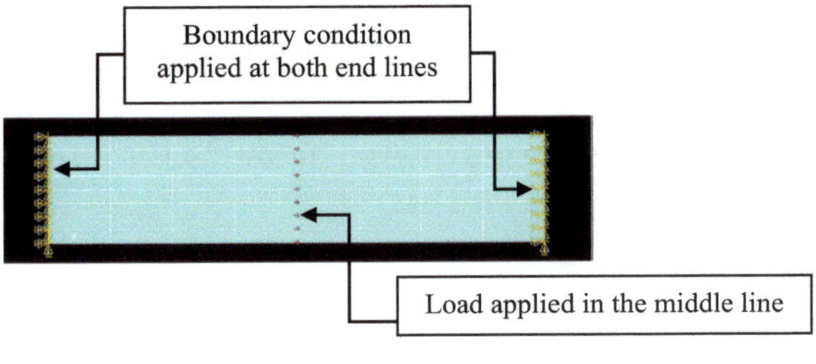

Fig. 1 Meshing element of the modelled specimen with boundary conditions

stress, ultimate compression stress, and shear stress in the standard ASTM material properties were inserted in the failure criteria which are used to determine the maximum stress of the material plate. The specimen's input data is listed in Table 1. The process of sectioning involves the orientation of the angle scheme for each layer of the laminate plate which includes the angle ply and the angle symmetric. The maximum stress value can be determined by using the interpolating technique based on failure index SMX = 1. Based on the result, the graph was plotted to show the behaviour of the materials.

Table 1 Input data of material properties of fibre composite laminates for pure and 25 wt.% nanosilica used for simulation in ANSYS [9, 10]

Material properties	APFRP				NS-25APFRP			GFRP
	10 vol.%	15 vol.%	20 vol.%	25 vol.%	5 wt.%	13 wt.%	25 wt.%	
Longitudinal modulus, E_1(GPa)	2.96	3.78	3.844	4.008	4.119	3.756	3.205	6.47
Transverse modulus, E_2(GPa)	0.296	0.378	0.3844	0.401	0.412	0.376	0.3205	2.16
Shear modulus, G_{12} (GPa)	1.85	2.3625	2.4025	2.505	2.574	2.3475	2.003	2.588
Poisson's ratio, ν_{12}	0.25	0.25	0.25	0.25	0.25	0.25	0.25	0.25
Shear strength, S_{12} (MPa)	31.637	34.19	37.901	39.635	41.086	37.214	22.911	41.00
Tensile strength, X_T (MPa)	54.83	59.257	65.686	68.692	71.206	64.496	39.707	342.56
Tensile strength, Y_T (MPa)	5.483	5.9257	6.5686	6.8692	7.1206	6.4496	3.9707	9.25
Compressive strength, X_C(MPa)	88.518	92.459	93.995	99.327	102.053	93.336	70.786	77.29
Compressive strength, Y_C(MPa)	8.8518	9.2459	9.3995	9.9327	10.2053	9.3336	7.0786	10.31

3 Results and Discussion

This section presents the results and discussion obtained from the simulation by ANSYS under bending loading. The effect of *Arenga Pinnata* fibre volume content and nanosilica content on the composites was also analysed. This study focused on the maximum stress distribution and maximum deflection of APFRP composites that were modelled with four different volume contents of APFRP (10, 15, 20 and 25 vol.%). Then the composites were modified with three different weight percentages of nanosilica addition (5, 13, and 25 wt.%). Finally, the results were compared with Glass FRP.

3.1 Effect of Arenga Pinnata Fibre Volume Content to the Maximum Bending Stress and Maximum Deflection at Failure Load Condition of Different Fibre Ply Orientations

Table 2 presents the results of maximum bending stress of four different volume contents of *Arenga Pinnata* FRP (10, 15, 20 and 25 vol.%) from ANSYS simulations. The table shows the value of maximum bending stress and maximum deflection with 0°, 15°, 30°, 45°, 60°, 75°, and 90° fibre ply orientation. Overall, the highest value of maximum bending stress was obtained in 25 vol.% of *Arenga Pinnata* FRP while the lowest was in 10 vol.% of *Arenga Pinnata* FRP. From the table, the value of maximum bending stress at 0° fibre ply orientation for 10APFRP was 58.5 MPa, while 15APFRP was 63.2 MPa. Compared with 20APFRP and 25APFRP, the values were 70.1 MPa and 73.3 MPa, respectively. From the results, it was shown that a higher volume content of *Arenga Pinnata* to the FRP composites increased the maximum bending stress of FRP. It was also observed that the maximum deflection of the APFRP increased when the maximum bending stress was increased.

Table 2 Result of maximum bending stress and maximum deflection of 10 vol.% APFRP, 15 vol.% APFRP, 20 vol.% APFRP, and 25 vol.% APFRP

Fibre ply orientation	10APFRP		15APFRP		20APFRP		25APFRP	
	(MPa)	(mm)	(MPa)	(mm)	(MPa)	(mm)	(MPa)	(mm)
$[0]_8$	58.5	4.59	63.2	3.88	70.1	4.23	73.3	4.25
$[+15/-15/0/0]_{2s}$	69.3	4.807	74.9	4.068	83	4.435	86.8	4.448
$[+30/-30/0/0]_{2s}$	71	4.818	76.7	4.078	85	4.445	88.9	4.458
$[+45/-45/0/0]_{2s}$	72.4	4.807	78.2	4.068	86.8	4.434	90.7	4.447
$[+60/-60/0/0]_{2s}$	57.5	4.599	62.1	3.892	80.3	4.184	72	4.255
$[+75/-75/0/0]_{2s}$	32.7	4.571	35.3	3.869	39.2	4.217	40.9	4.23
$[+90/-90/0/0]_{2s}$	7.5	4.55	8.11	3.85	8.99	4.2	9.4	4.21

Figure 2 illustrates the deformation contour of failure load conditions at 0° and 90° fibre ply orientation for 25 vol.% of the *Arenga Pinnata* FRP composite. The other three materials (10, 15, and 20 vol.%) also showed the same trend but with different values. When a higher force was applied, the maximum stress was also higher. From the result shown in the modelling, the red colour referred to the highest deflection that occurred at the middle of the specimen and blue at both ends. This is because the force was applied at the middle for the bending test and the set up was fixed at both ends of the specimen.

Based on the graph shown in Fig. 3, the trendline showed that the curve increased from 0° to 45° fibre ply orientation, then decreased from 60° to 90°. It also showed that when the angle of fibre orientation was 45°, it had the highest maximum bending stress compared to the other fibre ply orientation and the lowest was at 0°. The value maximum bending stress was at 45° fibre ply orientation for 10APFRP, 15APFRP, 20APFRP, and 25APFRP are 72.4 MPa, 78.2 MPa, 86.8 MPa, and 90.7 MPa, respectively. From the results, the load applied at 45° fibre ply orientation provided the highest stress as it gave the maximum force to the specimen better than other angles. From the results, it was apparent that 25 vol.% of *Arenga Pinnata* provided the highest bending stress compared to 10, 15, and 20 vol.%.

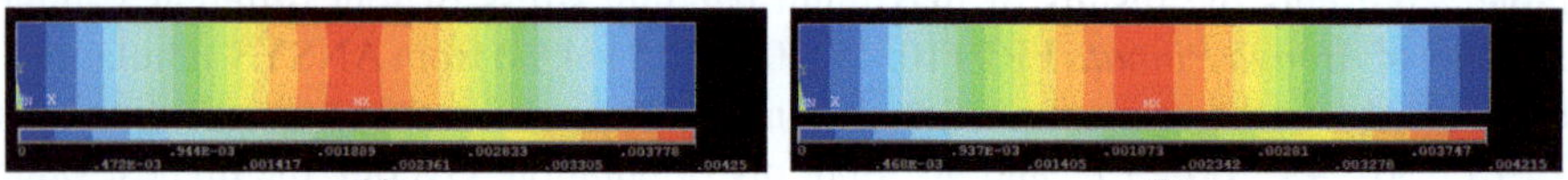

(a)
Failure load = 133.78 N
Maximum deflection = 4.25 mm

(b)
Failure load = 28.65 N
Maximum deflection = 4.21 mm

Fig. 2 Deformation contour at failure load condition of 25 vol.% of APFRP for **a** 0° fibre ply orientation and **b** 90° fibre ply orientation

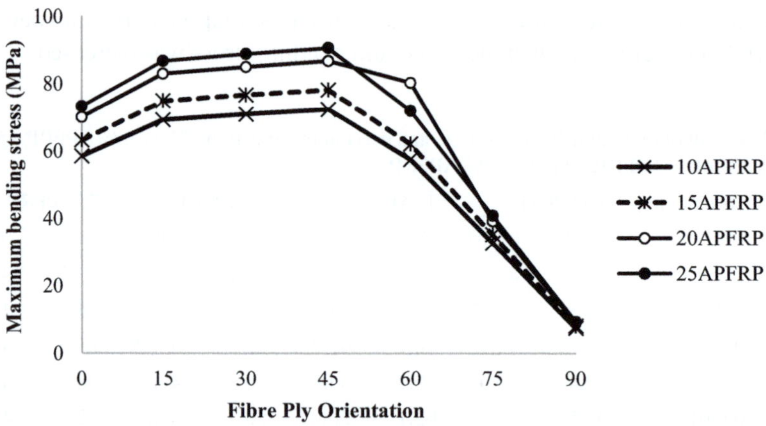

Fig. 3 Maximum bending stress curve at different fibre ply orientations for four different volume contents of *Arenga Pinnata* FRP (10, 15, 20, and 25 vol.%)

3.2 *Effect of Nanosilica Addition on 25 Wt.% APFRP Composites to the Maximum Bending Stress and Maximum Deflection at Failure Load Condition of Different Fibre Ply Orientations*

Table 3 presents the results of maximum bending stress of 25% volume content of *Arenga Pinnata* FRP with the addition of three different percentages of nanosilica content (5, 13, and 25 wt.%). The table shows the value of maximum bending stress and maximum deflection with 0°, 15°, 30°, 45°, 60°, 75°, and 90° fibre ply orientation. The highest value of the maximum bending stress was obtained in 5 wt.% nanosilica added to the 25APFRP, while the lowest was 25 wt.% nanosilica added to the 25APFRP. From the table, the maximum bending stress for 5 wt.% nanosilica content of 25APFRP at 0° fibre ply orientation was 76.0 MPa, whereas 68.8 MPa and 42.4 MPa for 13 wt.% and 25 wt.%, respectively. From the results, it showed that higher nanosilica addition to the *Arenga Pinnata* FRP decreased the value of the bending stress of APFRP. Similarly, it was also observed that the maximum deflection of the APFRP increased when the maximum bending stress was increased.

Figure 4 illustrates the maximum stress distribution at failure load conditions of 0° and 90° fibre ply orientation for 5 wt.% nanosilica content of the *Arenga Pinnata* FRP composite. 13 and 25 wt.% nanosilica content of the *Arenga Pinnata* FRP composite also showed the same trend but with different values. The maximum stress increased as the applied load was increased. Likewise, the red colour in the modelling results referred to the highest deflection that occurred in the middle of the specimen while the blue colour was shown at both ends. It is because the force was applied in the middle for the bending test and the set-up was fixed at both ends of the specimen.

Figure 5 shows the trendline of the maximum bending stress versus fibre ply orientation for 5, 13 and 25 wt.% of 25APFRP. It showed that the curve increased from 0° to 45° fibre ply orientation then decreased from 60° to 90°. Similarly, the highest maximum bending stress was at 45° fibre ply orientation while the lowest was at 0°. The value maximum bending stress was at 45° fibre ply orientation for

Table 3 Result of maximum bending stress and maximum deflection of nanosilica addition to the 25 vol.% APFRP (5, 13, and 25 wt.%)

Fibre ply orientation	5NS-25APFRP		13NS-25APFRP		25NS-25APFRP	
	(MPa)	(mm)	(MPa)	(mm)	(MPa)	(mm)
$[0]_8$	76.0	4.28	68.8	4.25	42.4	3.07
$[+15/-15/0/0]_{2s}$	90.0	4.486	81.5	4.456	50.2	3.215
$[+30/-30/0/0]_{2s}$	92.3	4.497	83.5	4.467	51.4	3.223
$[+45/-45/0/0]_{2s}$	94.1	4.486	90.2	4.543	52.5	3.215
$[+60/-60/0/0]_{2s}$	74.7	4.292	67.6	4.263	41.6	3.076
$[+75/-75/0/0]_{2s}$	42.4	4.266	38.4	4.238	23.7	3.057
$[+90/-90/0/0]_{2s}$	9.74	4.25	8.83	4.22	5.43	3.04

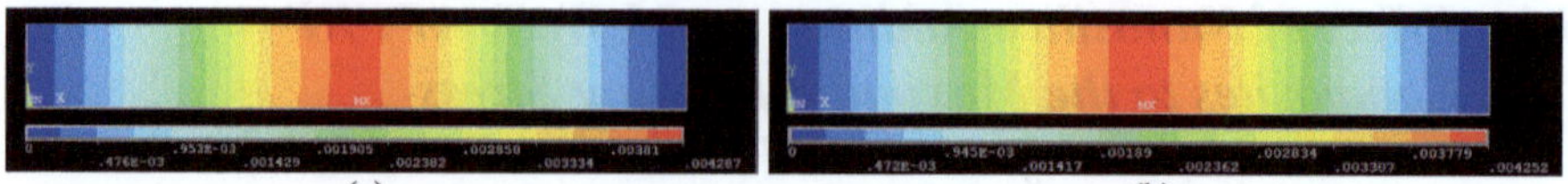

(a) Failure load = 138.68 N
Maximum deflection = 4.28 mm

(b) Failure load = 29.69 N
Maximum deflection = 4.25 mm

Fig. 4 Deformation contour at failure load condition of 5 wt.% of 25APFRP for **a** $[0]_8$ fibre ply orientation and **b** $[+90/-90/0/0]_{2s}$ fibre ply orientation

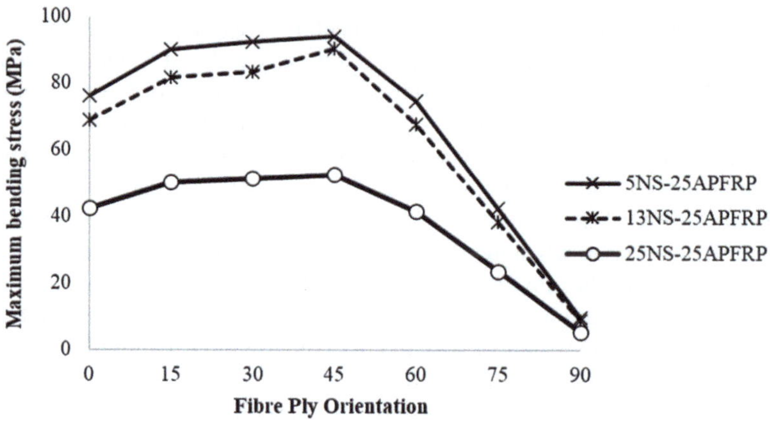

Fig. 5 Maximum bending stress curve at different fibre ply orientations for different nanosilica in addition to the 25 vol.% APFRP (5, 13 and 25 wt.%)

5NS-25APFRP, 13NS-25APFRP, and 25NS-25APFRP are 94.1 MPa, 90.2 MPa, and 52.5 MPa respectively. From the results, the load applied at 45° fibre ply orientation provided the highest stress as it gave better maximum force to the specimen than other angles.

3.3 Maximum Bending Stress and Maximum Deflection of 25APFRP and 5NS-25APFRP Composites Compared to GFRP Composites

The results of maximum bending stress and maximum deflection of 25 vol.% APFRP, 5 wt.% nanosilica addition to 25APFRP were then compared to Glass FRP composites as summarized in Table 4. The values for both 25APFRP and 5NS-APFRP were 90.7 MPa and 94.1 MPa, respectively. It showed that the highest maximum bending stress and maximum deflection occurred at 45° of fibre ply orientation while the lowest was at 90° fibre ply orientation. On the other hand, Glass FRP had the highest maximum bending stress at 0° fibre ply orientation and the lowest at 90° fibre ply

Table 4 Result of maximum bending stress and maximum deflection of 25APFRP, 5NS-25APFRP, and GFRP composites

Fibre ply orientation	25APFRP		5NS-25APFRP		Glass FRP	
	(MPa)	(mm)	(MPa)	(mm)	(MPa)	(mm)
$[0]_8$	73.3	4.25	76.0	4.28	80.8	2.948
$[+15/-15/0/0]_{2s}$	86.8	4.448	90.0	4.486	76.6	2.758
$[+30/-30/0/0]_{2s}$	88.9	4.458	92.3	4.497	51.4	1.789
$[+45/-45/0/0]_{2s}$	90.7	4.447	94.1	4.486	40.4	1.437
$[+60/-60/0/0]_{2s}$	72	4.255	74.7	4.292	28.3	1.212
$[+75/-75/0/0]_{2s}$	40.9	4.23	42.4	4.266	17.4	1.093
$[+90/-90/0/0]_{2s}$	9.4	4.21	9.74	4.25	10.2	1.059

orientation. Nevertheless, by comparing these three composites, the results showed that Glass FRP had the highest value of maximum bending stress but with low maximum deflection. Table 5 presents the comparison of the modelling of maximum stress distribution of failure condition at 0° fibre ply orientation. 25 vol.% of *Arenga Pinnata* FRP, 5 wt.% nanosilica content of 25APFRP composites and Glass FRP composites. These three composites showed the same trend but with different values.

3.4 Comparison of Maximum Bending Stress Curve for 25 Vol.% of APFRP, 5 Wt.% Nanosilica of 25APFRP and Glass FRP

Based on the graph shown in Fig. 6, the graph increased from 0° until 45° of fibre ply orientation and decreased to 90° for both 25 vol.% APFRP and 5 wt.% of 25APFRP composites. However, the graph continuously decreased from 0° to 90° for Glass FRP. Comparing Glass FRP to 25 vol.% APFRP and 5 wt.% of 25APFRP composites, the deflection for GFRP was much less than 25APFRP and 5NS-25APFRP because of its inability to behave in a plastic way.

4 Conclusion

This study represented the maximum bending stress and deformation analysis of the *Arenga Pinnata* fibre composite laminate and the effect of nanosilica addition to the APFRP by using the ANSYS software. From this study, it was found that the 45° unidirectional laminate exhibited the highest flexural strength when compared to all off-axis laminates. The maximum bending stress of 25APFRP was 90.7 MPa with a maximum deflection of 4.447 mm. The presence of 5 wt.% nanosilica improved the

Table 5 Displacement contour for 25 vol.% APFRP, 5 wt.% nanosilica content of 25APFRP composites and Glass FRP composites at 0° fibre ply orientation

FRP	Maximum deflection at z-direction	Maximum deflection at xy-direction
25APFRP	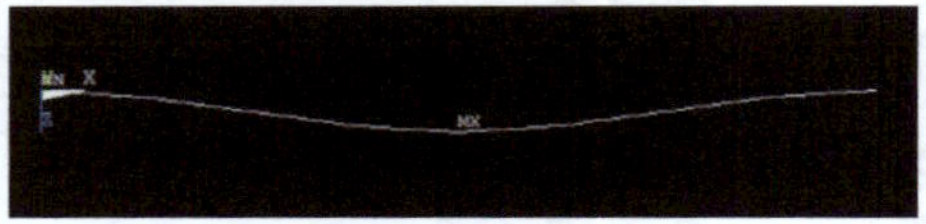	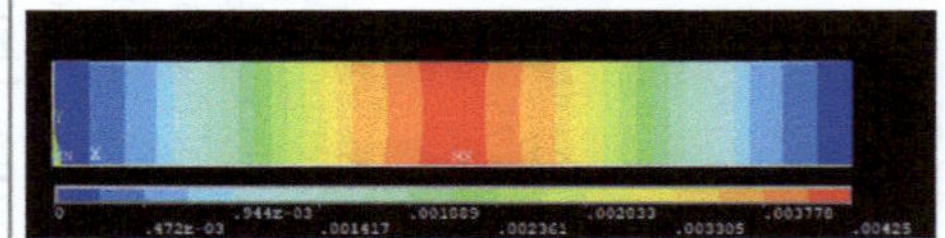 Maximum deflection = 4.25 mm
5NS-25APFRP	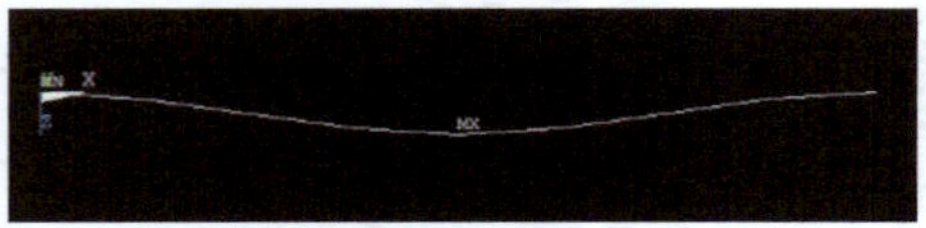	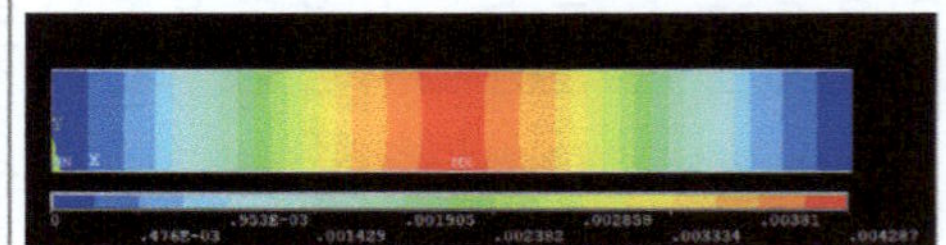 Maximum deflection = 4.28 mm
GFRP	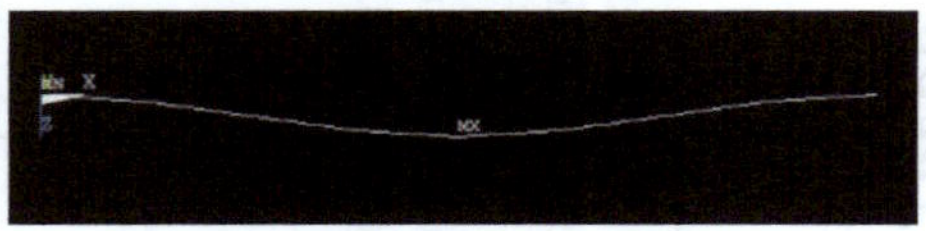	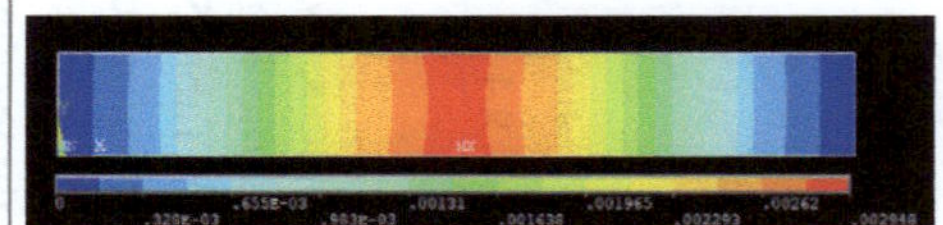 Maximum deflection = 2.948 mm

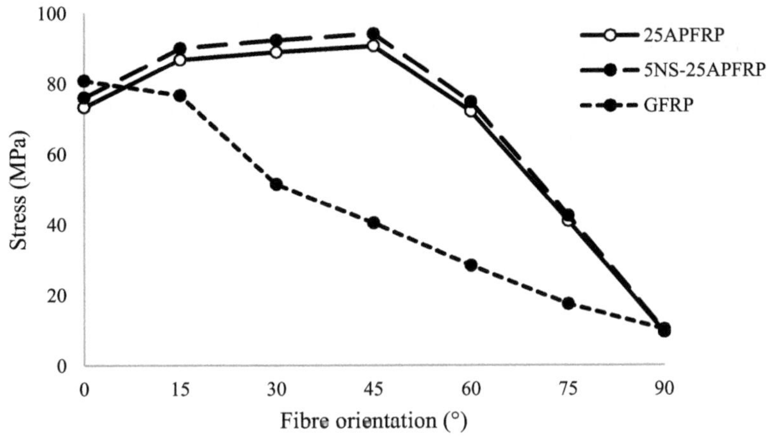

Fig. 6 Maximum bending stress of 25 vol.% of APFRP, 5 wt.% nanosilica of 25APFRP and Glass FRP

bending properties of *Arenga Pinnata* with a maximum bending stress of 94.1 MPa and 4.486 mm maximum deflection. It can be concluded that FRP composites made of *Arenga Pinnata* plant-based FRP composites have high flexural properties when compared to conventional Glass FRP composites.

Acknowledgements The authors gratefully appreciate the financial support from Universiti Teknologi MARA (UiTM) and the Ministry of Higher Education Malaysia. The research was performed at the School of Mechanical Engineering UiTM Malaysia under the support of Geran Sanjung Sarjana (GSS) Grant No: 600-RMC/GSS 5/3 (076/2022).

References

1. Ishak MR, Sapuan SM, Leman Z, Rahman MZA, Anwar UMK, Siregar JP (2013) Sugar palm (Arenga pinnata): its fibres, polymers and composites. Carbohydr Polym 91(2):699–710
2. Bhongade AS, Kumar Thakur K, Radkar DB (2020) Effect of fillers in glass matrix composite material suitable for light weight and high thermal strength applications. Mater Today Proc 38:2217–2221
3. Wang R-M, Zheng S-R, Zheng Y-P (2011) Introduction to polymer matrix composites. Polym Matrix Compos Technol 1–548
4. Bisanda ETN (1993) The manufacture of roofing panels from sisal fibre reinforced composites. J Mater Process Tech 38(1–2):369–379
5. Gupta N, Doddamani M (2018) Polymer matrix composites. Jom 70(7):1282–1283
6. Ogin SL, Brøndsted P, Zangenberg J (2016) Composite materials: constituents, architecture, and generic damage. Elsevier
7. Samuel OD, Agbo S, Adekanye TA (2012) Assessing mechanical properties of natural fibre reinforced composites for engineering applications. J Miner Mater Charact Eng 11(08):780–784

8. Eric Greene Associates (1999) Marine Composites-page 115, p 377
9. Khalid MFS (2017) Mechanical properties of treated arenga pinnata fibre reinforced nanomodified epoxy composite. Universiti Teknologi MARA
10. Hashim UR (2020) Mechanical properties of hybrid graphene filled basalt fibre reinforced. Universiti Teknologi MARA

Palm Byproducts Cementitious Composites

Physico-Mechanical Properties and Weathering Performance of Coconut Husk Fibre-Reinforced Composite Roofing Tiles Produced with Selected Cement Admixtures

Anthony O. Adeniji, Abel O. Olorunnisola, and Holmer Jr Savastano

Abstract Roofing constitutes a major cost item in building construction. The relatively high cost of conventional roofing materials is a factor militating against affordable housing provision in sub-Saharan Africa. This study examined the physico-mechanical properties and weathering performance of coconut husk fibre-reinforced composite roofing tiles produced with selected cement admixtures. Corrugated ($40 \times 30 \times 0.6$ cm^3) and flat ($16 \times 30 \times 0.6$ cm^3) roofing tiles were produced using Rice Husk Ash (ASH), Chicken Eggshell Ash (CESA) and Calcium Carbide Waste (CCW) as a partial replacement for Portland Limestone Cement (PLC). Formulations used were 4% coconut fibres by mass of cement, 0.4 for cement–water ratio and 1:2 for cement–sand. Cement admixtures were formulated at 90%PLC + 10%RHA; 90%PLC + 10%CESA; and 95%PLC + 5%CCW. All these were thoroughly mixed together and poured into a mould and covered with a polythene sheet. Composites were demoulded after 24 h, immersed in CO_2-injected water inside a controlled chamber for 4 min at 5.5 MPa and damp cured for 21 days. Three replicate samples were tested for density, Water Absorption (WA), Thickness Swelling (TS), Modulus of Elasticity (MOE) and Modulus of Rupture (MOR). Corrugated and flat tiles were installed on a pitched roof structure and exposed to natural tropical weathering conditions for 24 months in Ibadan, Nigeria. Weather data were collected and triplicate samples of the installed roofing tiles were tested for density, WA and TS at 90-day intervals. Composite density (1.9–2.0 g/cm^3), WA (7.5–8.7%) and TS (7.8–11.1%) were relatively high. Partial replacement of cement with RHA, CCW and CESA slightly reduced the density and TS but significantly increased the WA ($p \leq 0.05$). The MOE (1.5–2.8 GPa) and MOR (1.4–5.0 MPa) were relatively low. However, RHA and CCW significantly increased MOE and MOR. Temperature, relative humidity and rainfall values for the weathering test duration were 28.8- 33.4 °C, 13.0–88.5% and 5.1–16.4 mm respectively. All installed tiles exhibited a minimal reduction in

A. O. Adeniji (✉) · A. O. Olorunnisola
Department of Wood Products Engineering, University of Ibadan, Ibadan, Nigeria
e-mail: toneedenk@gmail.com

H. Jr Savastano
Faculty of Biosystem Engineering, University of Sao Paulo, Pirassununga, Brazil
e-mail: holmersj@usp.br

M. Jawaid et al. (eds.), *Proceedings of 2nd World Conference on Byproducts of Palms and Their Applications*, Springer Proceedings in Materials 19,
https://doi.org/10.1007/978-981-19-6195-3_9

density (<0.6 g/cm^3), WA (3.0–4.6%) and TS (approximately 3%). Flat roofing tiles containing 10%RHA gave the best overall performance.

Keywords Coconut husk · Rice husk ash · Eggshell ash · Carbide waste · Roofing tiles

1 Introduction

Roofing constitutes a major cost of building construction globally. However, since most conventional building materials including asphalt shingles, steel, zinc, aluminium roofing sheets, etc., are imported into Nigeria and many other sub-Saharan African countries, the cost is typically influenced by the prevailing exchange rate and may be quite unpredictable, depending on the stability of the national economy in question. One way of addressing this challenge is by promoting the development of equally durable non-conventional roofing materials based on locally available raw materials such as natural fibres. A notable candidate material for non-conventional roofing production is the fibre-reinforced cement composite. Fibre-reinforced composites can be manufactured in different shapes and sizes, they also possess excellent fire and water resistance. Their relatively low thermal conductivity is another positive attribute thus recommending them for roofing sheet production [1, 2]. However, the sometimes relatively low adhesion between the fibre and the matrix tends to reduce their mechanical performance [3, 4]. Also, the mineralization of cellulose fibre in alkali cement is a major limit in a fibre-reinforced composite of the fibre-reinforced cement [5, 6]. The best solution to the mineralization of cellulose fibre is to use a less belligerent matrix by bridging composite ductility, maintaining quality and fibre strength.

The choice of fibres for composite production is often based on availability, processing cost and compatibility with Portland cement. Fibrous materials suitable for cement-bonded fibre-reinforced composite roofing tile production in Nigeria include bamboo, rattan cane, sugar cane bagasse, raffia palm and luffa (*Luffa cylindrica*), Cissus (*Cissus populnea*), and coconut husk, among others [7]. However, coconut husk has relatively stronger fibres and is easier to process than most of the aforementioned fibre sources [8, 9]. There are about three million coconut palm trees producing approximately 70 million coconuts annually in Nigeria. The average mature coconut weighs 680 g, about 42% of which is made up of the husk. The husk fibres are largely treated as waste [10].

Environmental concerns associated with CO_2 emission in cement manufacture coupled with its rising cost in Nigeria and similar developing countries constitute a challenge to fibre-reinforced cement composite roofing tile manufacture. Alternatively, cement provides an excellent technical option and has the potential to make a significant contribution to the manufacture of affordable cement-bonded composite roofing tiles. Pozzolanas are materials containing reactive silica and/ or alumina which on their own have little or no binding property but when mixed with cement in the presence of water, will set and harden like cement. The addition of pozzolana in cement-based products has two major advantages. Firstly, the properties of the cement are improved, and secondly, as the costs of a pozzolana are usually low and well below that of cement; the overall cost is significantly reduced assuming that pozzolana does not have to be transported too far. Pozzolanas can be divided into two groups: natural pozzolanas such as volcanic ash and diatomite, and artificial pozzolanas such as calcined clays, pulverized fuel ash and ash from burnt agricultural wastes. Many plant ashes have a high silica content and are therefore suitable as a pozzolana. In recent years, considerable research has gone into identifying plant wastes whose ashes produce good pozzolanas and which are available in exploitable quantities. Rice husk, a waste product of rice milling has been shown to have the greatest potential, because of its availability in large quantities in many parts of the world. Also, it produces a good quality quantity of ash when burnt and the ash content contains about 90% silica which makes it an excellent pozzolana. The relatively high silica content of rice husk ash (RHA) results in the formation of Calcium Silicate Hydrate (CSH) which contributes positively to strength development during curing. One of the most common pozzolanic agricultural wastes is the rice husk ash [11].

Another source of cementitious materials that can be used as a partial replacement for cement in composite roofing tile production is a wide array of calcium carbonate containing materials, Eggshells are predominantly calcium carbonate and can be used as a cement admixture material. Ashes of eggshells are largely made up of calcium oxide. Calcium carbide waste is another major source of calcium oxide. A number of studies have explored the use of eggshell ash and calcium carbide waste as partial replacements for cement in composite roofing tile production [10]. It has been shown that the higher the percentage of eggshell ash in concrete, the faster the rate of setting of cement [12]. Also, the conventional method of curing cement-based roofing materials is either by soaking them in water or placing them under wet clothes for periods lasting up to three weeks. However, it has been shown that the production of natural fibre cement composites can benefit from accelerated curing of the cement-based matrix in a carbon dioxide rich environment. Early exposure of cement binders to carbon dioxide (CO_2) can substantially reduce their setting and hardening times [5, 13–15].

Roofing tiles and/or sheets are expected to perform in outdoor conditions. However, despite a number of experimental studies on the production of cement-bonded fibre-reinforced roofing tiles and accelerated ageing tests [6, 16], limited studies have been reported on the real-life weathering performance of such products exposed to outdoor conditions. The aim of this study, therefore, was to determine weathering effects over a period of 24 months on physico-chemical properties of coconut fibre-reinforced roofing composite roofing tiles cured in a CO_2-rich environment.

2 Materials and Methods

2.1 Methodology

Coconut husk was collected from a mill, and fibres were removed from the husk, beaten to remove particles and chopped into different sizes of an average length of 15 mm. Chicken eggshells were collected and sun-dried for 2 days to remove the moisture and grinded into finest particles and sieved with a 75 μm sieve size. Rice husk was incinerated at 800^0C for ash production. Calcium carbide was sun-dried and sieved with a 75 μm. The proximate compositions of coconut fibres and the specific gravity of RHA, CESA and CCW were determined using ASTM C 618–05 and ASTM C 948–81. Cement–water ratio was fixed at 1:0.4, the cement–sand ratio was fixed at 1:2 while the coconut fibre was fixed at 4% (mass of cement) for all the formulations including control at 100% cement. The formulations for admixture were 90% LPC + 10%RHA, 90% LPC + 10%ESA and 95%CCW + 5%CCW, respectively. All the formulations were mixed thoroughly for fibre-reinforced composite matrix production.

The mixtures were poured in the mould of dimensions 40 × 30 × 0.6 cm^3 for corrugated and 16 × 30 × 0.6 cm^3 for flat roofing tiles, and spread uniformly to a thickness of 6 mm. The composite was vibrated for 60 s to reduce the porosity and smoothened with a spatula. Samples for corrugation were transferred to a corrugated mould and sealed in a plastic bag and flat tiles also were sealed in a plastic bag and were both left for 24 h at room temperature. The control and other formulations were wet cured in a water–CO_2 chamber for 4 min at 5.5 MPa and samples were left in the chamber for 24 h for good absorption of CO_2. Three replicate samples were produced for both flat and corrugated tiles. Bulk density, water absorption, thickness swelling, moduli of elasticity and moduli of rupture were tested on both composite roofing types tiles using ASTM D 1037 (1991), ASTM D 1037/89 (1991), ASTM D3043 00(2011)). The effects of weathering on the properties of composite roofing tiles were determined over a period of 2 years. Data were analysed using descriptive and ANOVA at $\alpha_{0.05}$.

3 Results and Discussions

3.1 Proximate Analysis and Specific Gravity

Table 1 shows the proximate composition of coconut fibre while Table 2 shows the specific gravity of coconut fibre, rice husk ash, chicken eggshell ash, calcium carbide waste and Portland limestone cement.

The crude fibre content of 7.7% reported for coconut fibre was lower than 46.1% for *Corcurbita moschata,* a vegetable species [22]. The high percentage of cellulose

Table 1 Proximate analysis of coconut fibre

Parameter	Composition (%)
Crude fibre	7.7
Hemicellulose	15–28
Cellulose	35–60
Lignin	25–48
Carbonhydrate	36.7

Table 2 Specific gravity of coconut fibre, RHA, CESA, CCW and PLC

Materials	Specific gravity
Coconut fibre	1.38
Rice husk ash (RHA)	2.19
Chicken eggshell ash (CESA)	2.49
Calcium carbide waste (CCW)	2.29
Portland limestone cement	3.08

content (35–60%) in the coconut fibre has a positive influence on the bending strength when used for composite production. Although, the carbohydrate (36.7%) content suggests that coconut fibre is susceptible to fungi attack, this can be addressed by pre-treating the fibre before being used in composite production [23]. The specific gravity is a measure of the index of mineral content and the observed specific gravity of the coconut fibre (1.38) was the lowest among the materials tested. The specific gravities of calcium carbide waste (2.3) and rice husk ash (2.2) obtained were lower than that of limestone Portland cement (3.08) and show that they would lower the density of composites if mixed with cement in partial replacement.

3.2 Physical and Mechanical Properties of the Roofing Composite Tiles

The average bulk densities (BD), water absorption (WA), thickness swelling (TS), moduli of elasticity (MOE) and moduli of rupture (MOR) of the fibre-reinforced composite roofing tile samples are presented in Table 3.

3.2.1 Bulk Density

As shown in Table 3 the bulk density ranged between 1.8 and 2.0 gcm^{-3}. The differences in bulk densities of composite tiles might be a result of the density of raw materials. Samples containing rice husk ash had the lowest density (1.8 gcm^{-3}), while samples containing eggshell ash had the highest (2.0 gcm^{-3}), though the observed

Table 3 Physical and mechanical properties of fibre-reinforced composite roofing tiles

Sample	BD	WA	TS	MOE	MOR
Formulation	(gcm^{-3})	(%)	(%)	(GPa)	(MPa)
Control	1.90^{A}	7.50^{A}	11.1^{A}	2.40^{A}	3.03^{A}
10% rice husk ash	1.80^{A}	8.70^{B}	10.5^{B}	2.60^{B}	5.01^{B}
10% eggshell ash	2.00^{A}	8.10^{C}	9.80^{C}	1.50^{C}	1.44^{C}
5% calcium carbide waste	1.90^{A}	7.70^{D}	7.80^{D}	2.80^{D}	4.05^{D}

Mean values with the same letters (A, B, C or D) are not statistically different

differences were not significant ($P \geq 0.05$). The values are similar to those reported for sisal roofing tiles [5].

3.2.2 Water Absorption

The water absorption (WA) varied significantly between 7.5 and 8.7%. These values were higher than 1.0–3.9% reported for coconut husk fibre-reinforced roofing tiles [10] but lower than 23.9% for composite produced from *Eucalyptus grandis* [24], bagasse (4.8–16.4%), aspen (24.0–58.0%) and larch (36.0–56.0%) [25–27]. Partial replacement of cement with rice husk ash, calcium carbide waste and eggshell led to a general increase in water absorption by the composites. The samples containing calcium carbide had the lowest WA (7.7%), while samples containing rice husk ash had the highest water absorption (8.7%). The relatively high WA of the samples containing rice husk ash might be attributable to the water affinity of the ash due to the relatively high silica oxide content [28]. In general, WA < 25% is an indication that the tiles are relatively stable in moist conditions and would be suitable for use in outdoor applications where they would be in constant exposure to humidity and moisture conditions.

3.2.3 Thickness Swelling

Thickness swelling is a measure of dimensional stability of fibre-reinforced cement-bonded composites. Table 3 shows that partial replacement of cement significantly reduced thickness swelling from 11.1% (for the control sample) to 10.5%, 9.8% and 7.8% for samples containing rice husk ash, eggshell ash and calcium carbide waste, respectively. However, the values were all higher than a range of 0.5–3.1% reported for coconut husk-cement composite [29] and 0.6–4.0% for cement-bonded rattan fibre composites [30]. The main contributory factor to thickness swelling cement-bonded composites, which might have accounted for the deviation of results obtained in this study from previous ones, has been identified as the compressive strain and stress developed during production [31].

3.2.4 Moduli of Elasticity and Rupture

The range of MOE of composites was between 1.5 and 2.8 GPa. The observed differences were significant ($P \leq 0.05$). In general, the MOEs were lower than values reported for composites made from *Eucalyptus grandis* [24]. Samples containing rice husk ash and calcium carbide waste performed better than the control. However, the average MOE of only the samples containing calcium carbide was close to 3.0 GPa prescribed by the International Standard Organization (ISO). The average Modulus of Rupture (MOR) ranged between 1.5 and 5.0 MPa. These values are comparable with those reported for composites reinforced with rattans (1.6–6.3 MPa) [31] and 3.8–6.2 MPa reported for composites reinforced with sisal fibre [32], but lower than the 15.6–22.2 MPa reported for composites reinforced with *Eucalyptus grandis* [24]. The failure pattern observed showed that composites were brittle. Again, samples containing rice husk ash and calcium carbide waste performed better than the control.

3.3 Effects of Weathering on Physical Properties of Corrugated and Flat Roofing Tiles

The effects of rainfall, relative humidity and temperature variation on the bulk density, water absorption and thickness swelling of both corrugated and flat roofing tiles over a period of 720 days of exposure i.e., two cycles of rainy and three cycles of dry seasons, is discussed below.

3.3.1 Weathering Effects on Bulk Densities of Composite Roofing Tiles

The effects of weathering on the bulk densities of the flat and corrugated roofing are shown in Figs. 2 and 3. There was a general reduction in the bulk density due to weathering in both flat and corrugated roofing tiles regardless of composition. There was an initial increase in the bulk density of all the roofing tiles after exposure to the first rainy season of 2017 which might be accounted for by the permeability and coconut fibre (Fig. 1). Leaching was also observed on the surface of all the roofing tiles, resulting in fibre exposure. However, the fibres were more exposed on the flat roofing tiles than on the corrugated ones. For the flat roofing tiles, the greatest reduction in bulk density of 0.528 gcm^{-3} was observed in the samples containing 10% rice husk ash while the least reduction of 0.26 gcm^{-3} occurred in the control sample. For the corrugated roofing tiles, the greatest reduction in bulk density (0.37 gcm^{-3}) was observed in the sample made of 10% eggshell ash while the least reduction (0.28 gcm^{-3}) occurred in the control sample.

The observed trend in the reduction of bulk density of the roofing tiles can possibly be attributed to leaching, particularly for the composite containing cement admixtures after 720 days of weather exposure. This trend is similar to the findings of the

Fig. 1 Flat and corrugated roofing tiles at 720 days of weathering

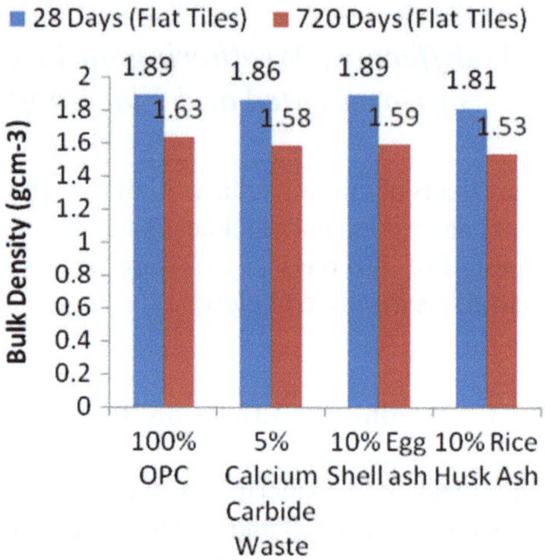

Fig. 2 Bulk densities of flat roofing tiles roofing at 720 days weathering

composite produced from sisal under natural weathering [33]. It can be assumed from the findings that the cementitious admixtures and corrugation had a great effect on the rate of leaching of the composite materials.

3.3.2 Weathering Effects on Water Absorption of Composite Roofing Tiles

The effects of weathering on water absorption of the flat and corrugated roofing tiles are shown in Figs. 4 and 5. There was a general reduction in the water absorption due to weathering, for all the roofing tiles (both flat and corrugated regardless of composition). However, for flat tiles, the greatest reduction in water absorption (4.6%) was

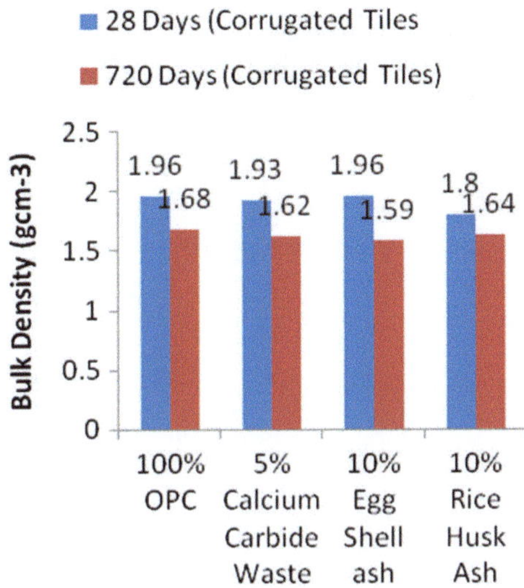

Fig. 3 Bulk densities of corrugated tiles at 720 days weathering

observed in the samples containing 10% rice husk ash while the least reduction was observed in samples made from 10% eggshell ash (3.0%). For the corrugated roofing tiles, the greatest reduction in water absorption (4.1%) was observed in samples made of 10% rice hush ash while the least reduction (2.1%) occurred in the control sample. The range of water absorption values recorded for all the formulations at the end of the weathering test was lower than 36.0% reported for composite roofing tiles produced with eucalyptus subjected to 13 months weathering test and 32.0% for sisal fibre reinforced composites subjected to 155 days weathering test [34]. Samples containing rice husk ash which exhibited the highest initial water absorption (7.9% for flat tiles, 8.4% for corrugated tiles), exhibited the least water absorption (3.3% for flat tiles, 4.3% for corrugated tiles) at the end of the weathering test. This implies that rice husk ash has a long-term water resistance potential that may even be superior to that of ordinary Portland cement.

3.3.3 Weathering Effects on Thickness Swelling of Composite Roofing Tiles

The effects of weathering on the thickness swelling of flat and corrugated roofing tiles are shown in Fig. 6 and 7. There was a general reduction in thickness swelling due to weathering, for all the roofing tiles (both flat and corrugated regardless of composition). However, the corrugated tiles exhibited more thickness swelling than the flat ones. For the flat tiles, the greatest reduction in thickness swelling was observed in the control sample (2.76%) while the least reduction in thickness swelling (1.83%) was observed in samples containing 5% calcium carbide waste. For the corrugated tiles, the greatest reduction in thickness swelling (2.47%) was observed in the sample made of 10% rice husk ash while the least reduction occurred in the control (1.09%).

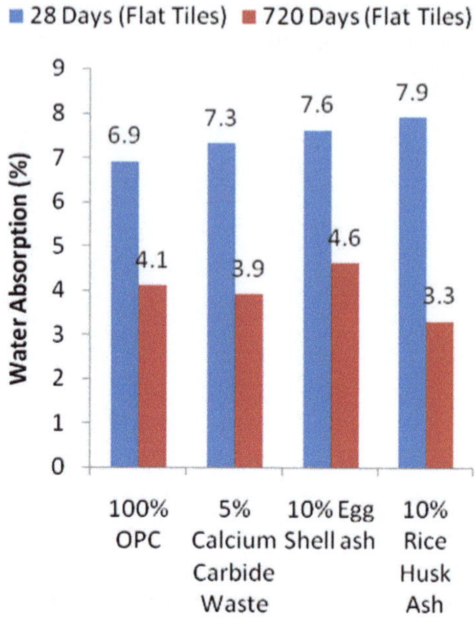

Fig. 4 Water absorption of flat roofing tiles at 720 days weathering

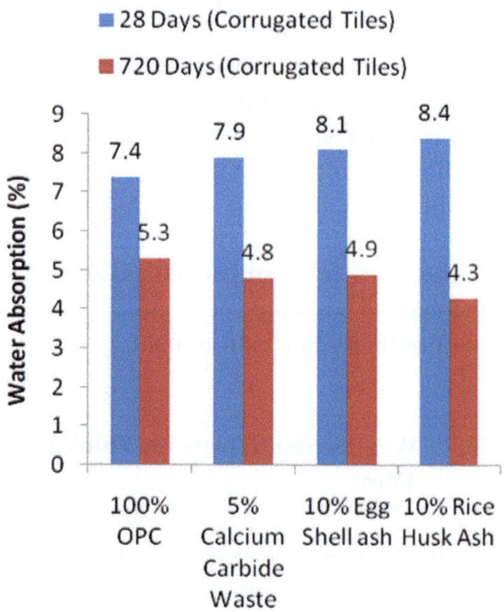

Fig. 5 Water absorption of corrugated roofing tiles at 720 days weathering

Again, these findings suggest that rice husk ash has a long-term ameliorating effect on the property of roofing tiles to swell on account of water absorption. There were no readily available data in literature to compare these findings.

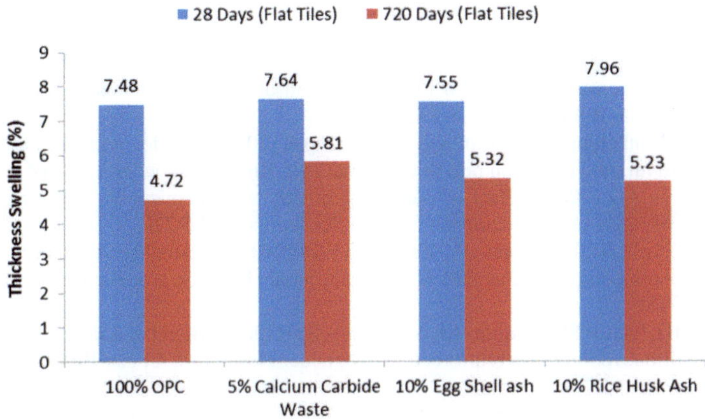

Fig. 6 Thickness swelling of flat tiles at 720 days weathering

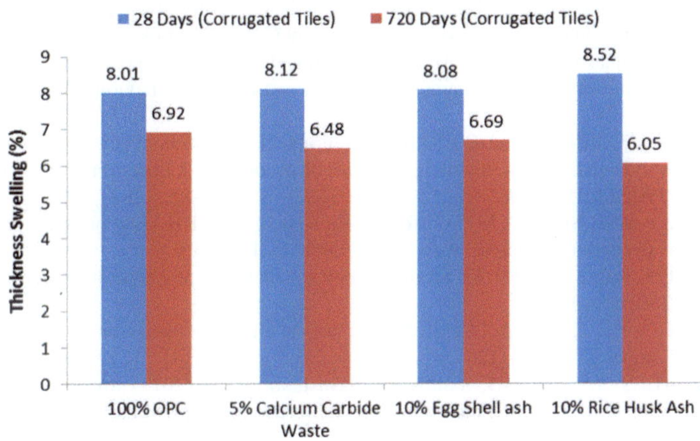

Fig. 7 Thickness swelling of corrugated tiles at 720 days weathering

4 Conclusion

The results of the experiment showed that rice husk ash and calcium carbide waste in particular can be used for partial replacement of cement in coconut fibre-reinforced cement composite roofing tiles for use in South-western Nigeria. Preliminary evidence suggests that the coconut fibre-reinforced composite roofing tiles could withstand long-term weathering conditions in tropical environments and that flat roofing tiles would be better than the corrugated ones. It is concluded that coconut fibre-reinforced composites can serve as a good alternative to conventional roofing materials in tropical Africa.

References

1. Khedar J, Charoenvai S, Hirunlabh J (2003) New insulating particle boards from durian peel and coconut coir. Build Environ 38:245–249
2. Khedari J, Nankongnab N, Hirunlabh J, Teekasap S (2004) New low-cost insulation particle boards from mixture of durian peel and coconut coir. Build Environ 39(1):59–65
3. Olorunnisola AO (2007) Effect of particle geometry and chemical accelerator on strength properties of rattan-cement composites. Afr J Sci Technol 8:22–27
4. Ikai S, Reichert JR, Rodrigues AV, Zampieri VA (2010) Asbestos-free technology with new high toughness polypropylene (PP) fibres in air-cured Hatschek process. Constr Build Mater 24(2):171–180
5. Tonoli GHD, Santos SF, Joaquim AP, Savastano (Jr) H (2010) Effect of accelerated carbonation on cementitious roofing tiles reinforced with lignocellulosic fibre. Constr Build Mater 24:193–201
6. Mohr BJ, Nanko H, Kurtis KE (2005) Durability of Kraft pulp fibre-cement composites to wet/dry cycling. Cement Concr Compos 27(4):435–448
7. Olorunnisola AO (2006) Strength and water absorption characteristics of cement-bonded particle board produced from coconut husk. J Civ Eng Res Pract 3(1):41–49
8. Ramakrishna G, Sundararajan T (2005) Studies on the durability of natural fibres and the effect of corroded fibres on the strength of mortar. Cement Concr Compos 27(5):575–582
9. Abdulahi S, Baharin J, Mazlee MN, Kamarudin H (2011) Composite cement reinforced coconut fibre: physical and mechanical properties and fracture behaviour. Aust J Basic Appl Sci 5(7):1228–1240
10. Olorunnisola AO, Adeniji AO (2019) Investigations on the effects of cement replacement and calcium chloride addition on selected properties of coconut husk fibre-reinforced roofing tiles. Mater Res Proc 11:253–259
11. Maisarah Ali MSF, Siti AS (2015) Effect of Rice Husk Ash (RHA) on physical property and mechanical strength of concrete. Adv Mater Res 1115:150–155
12. Mtallib MOA, Rabiu A (2009) Effect of eggshell ash on the setting time of the cement. Nigeria J Technol 28(2):29–38
13. Frias M, Gonis S (2013) Accelerated carbonation effect on behavior of tertiary Portland cements. Compos B 48:122–128
14. Almeida AEFS, Tonoli GHD, Santos SF, Savastano (Jr) H (2013) Improved durability of vegetable fiber reinforced cement composite subject to accelerated carbonation at early age. Cem Concr Compos 42:49–58
15. Pizzol VD, Mendes LM, Frezzatti L, Savastano (Jr) H, Tonoli GHD (2014) Effect of accelerated carbonation on the microstructure and physical properties of hybrid fiber–cement composites. Miner Eng 59:101–6
16. Aznizam A, Azman H, Ahmad-Fuad M (2005) Effects of accelerated weathering on the mechanical properties of oil palm empty fruit bunch filled UPVC composite. Iran Polym J 14(4):627–635
17. ASTM C 618 (2005) Standard Specification for coal fly ash and raw or calcined natural pozzolan for use as a mineral amixture in concrete, Philadephia, pp 310–313
18. ASTM C 948–81 (1981) Test Method for dry and wet bulk density, water absorption and apparent porosity of the section of glass-fibre reinforced concrete
19. ASTM D1037-12 (2012) Standard test methods for evaluating properties of wood-base fibre and particle panel materials, USA
20. ASTM D1037-99 (1999) Standard test method for evaluating properties of wood-base fibre and particle panel material, USA
21. ASTM D3043-00 (2000) Standard test method for structural panels in flexure
22. Hussain J, Rehman N, Khan A, Hamayun M, Hussain SM, Shinwari ZK (2010) Proximate and essential nutrients evaluation of selected vegetables species from Kohat Region, Pakistan. Pak J Bot 42(4):2847–2855

23. Asasutjarit C, Hirunlabh J, Charaenvai S, Zeghmati B, Cheul Shin U (2007) Development of coconut coir based lihht weight cement board. Constr Build Mater 21:277–288
24. Savastano (Jr) H, Warden PG, Coutts RSP (2003) Potential of alternative fibre cement as building materials for developing areas. Cement Concr Compos 31:232–243
25. Moslemi AA, Souza M, Geimer R (1994) Accelerated ageing of cement-bonded particle board. In: 4th international inorganic-bonded wood-fibre composite materials conference, Spokane, Washigton, USA, pp 83–83
26. Oyegade OA (2000) Thickness swelling components and water absorption of cement-bonded particle boards made from Gmelina wood, Bagasse and Coconut Husk. Niger J For 30(1–2):10–14
27. Ajayi B (2002) Investigation of the dimensional stability of cement-bonded Composite board fabricated from coffee husks. Niger J For 33(2):88–93
28. Narayan PS (2002) Use of rice husks for concrete moulding. Res Ind 20(3):142–150
29. Olorunnisola AO (2009) Effects of husk particle size and calcium chloride on strength and sorption properties of coconut husk-cement composites. Ind Crops Prod 29:495–501
30. Olorunnisola AO, Agrawal S (2015) Effects of NaOH concentration and fibre content on physic-mechanical properties of cement-bonded rattan fibre composites. Pro Ligno Int J Wood Eng 11(4):192–198
31. Badejo SO (1988) Effect of flake geometry on properties of cement-bonded particle board from mixed tropical hardwoods. Wood Sci Technol 22:357–370
32. Agopyan V, Savastano (Jr) H, John VM, Cincotto MA (2005) Development on vegetable fibre-cement based materials in Sao-Paulo, Brazil: an overview. Cement Concr Compos 27(5):527–536
33. Savastano (Jr) H, Warden PG, Coutts RSP (2001) Performance of low-cost vegetable fibre-cement composite under weathering. In: CIB world building congress, Wellington, New Zealand
34. Roma (Jr) LC, Martello LS, Savastano (Jr) H (2008) Evaluation of mechanical, physical and thermal performance of cement-based tiles reinforced with vegetable fibre. Constr Build Mater 22:668–674

Effects of Partial Replacement of Cement with Selected Polymers on Sorption and Mechanical Properties of Rattan Cane Fibre-Reinforced Composite Roofing Tiles

A. Ogundipe and Abel O. Olorunnisola

Abstract Environmental concerns about CO_2 emissions during Portland cement manufacture have drawn attention to the need to minimise cement consumption. Mixing cement with relatively cheap and readily available polymers is one way of addressing the challenge. This study, therefore, investigated the effects of polymer–cement admixtures on selected properties of rattan cane fibre-reinforced roofing tiles. Mature rattan (*Laccosperma secundiflorum*) canes were sun-dried and hammer-milled. The fibres were treated with dilute NaOH (10% w/v). Three polymeric materials were used in polymer–cement admixtures—natural rubber latex; *Cissus populnea* gel extracted from the stem of cissus plant; and acrylic emulsion latex, a synthetic paint. Triplicate samples of 160 × 50 × 6 mm composite tiles were produced with 3% fibre content using 0.5 cement/water ratio. Based on preliminary studies, cement was partially replaced (w/w) with 5,7.5, 10% natural rubber latex;10, 20%, 30% acrylic emulsion latex; and 10,20,30% *Cissus populnea* gel in different composite samples cured under wet conditions for 28 days. The Density, Water Absorption, Thickness Swelling, Apparent Porosity, Modulus of Elasticity, and Modulus of Rupture of the samples were determined using standard methods. Composite densities ranged between 0.86 and 1.23gc/m^3. All the three polymers, particularly natural rubber latex, significantly reduced the density ($p \leq 0.05$). Water Absorption ranged between 0.62 and 1.66%. Though *Cissus populnea* gel significantly increased Water Absorption, the values still fell within acceptable limits. Thickness Swelling (0.54–2.76%) was relatively low. Apparent Porosity (11.6–28.5%) was also relatively low except in samples containing acrylic emulsion latex. The Modulus of Elasticity (930-4649 N/mm^2) and Modulus of Rupture (1.12–4.58 N/mm^2) were within acceptable limits. However, natural rubber latex had negative, while *Cissus populnea* gel had positive effects on both strength properties with the gel increasing the Modulus of Elasticity and Modulus of Rupture by approximately 100% of the control sample values. It was concluded that relatively strong and dimensionally stable rattan fibre-reinforced roofing tiles can be produced with *Cissus populnea* gel–cement admixtures at cement replacement levels of up to 30%.

A. Ogundipe · A. O. Olorunnisola (✉)
Department of Wood Products Engineering, University of Ibadan, Ibadan, Nigeria
e-mail: abelolorunnisola@yahoo.com

M. Jawaid et al. (eds.), *Proceedings of 2nd World Conference on Byproducts of Palms and Their Applications*, Springer Proceedings in Materials 19,
https://doi.org/10.1007/978-981-19-6195-3_10

Keywords Rattan fibres · Polymer–cement · *Cissus populnea* · Roofing tiles

1 Introduction

The need to improve housing supplies in developing countries is great. Roofing is a fundamental requirement in building construction and constitutes a substantial fraction of the building cost in Nigeria and many other sub-Saharan countries [13]. Over the ages, several roofing materials have emerged including ceramic (clay-based), asphalt, concrete, asbestos, metallic (such as corrugated iron, steel, zinc- or aluminum-based) and composite materials (including fire-reinforced composites), among others. In many African countries, corrugated zinc roofing sheets are the least expensive type of metallic roofing materials now in general use in low-cost houses and other buildings. However, the roofing materials that last the longest are concrete, clay or slate tiles. These materials significantly outperform other natural products like wood shakes or any manufactured roofing materials including asphalt shingles and metal roofing.

Concrete tiles, sometimes called cement roofing tiles, are a mixture of Portland cement, sand, water, and sometimes various dyes (for colour). Due to the fluid nature prior to curing, concrete tiles are factory-moulded into a wide variety of shapes and textures to simulate wood shake shingles, clay tiles, or slate tiles. Portland cement is the most common type of cement in general use for making cementitious products. However, Portland cement manufacture has several negative environmental impacts including the release of CO_2 from raw materials; emissions of other airborne pollutants in the form of dust, gases, noise, and vibration; consumption of large quantities of fuel; and damage to quarrying sites. Another major challenge in many developing countries such as Nigeria is the rising cost of cement. For example, a recent market survey by the authors showed that the price of Portland cement in Nigeria has doubled from 40,000 Naira (US$ 80) to 80,000 Naira (US$160) per tonne in the last five years.

To address environmental and economic challenges associated with Portland cement consumption, it is expedient to identify suitable cementitious admixtures for partial replacement of Portland cement without compromising the physical and mechanical properties of the products obtained. Such admixtures include polymer–cement admixtures incorporating natural rubber latex, *Cissus populnea* gel (colloids of fermented Cissus plant stem) extracted from the stem of cissus plant; and acrylic emulsion latex [4, 11]. Natural rubber latex is made of sap harvested from rubber trees. The sap is whipped into a froth and baked into cozy latex layers. The latex is the polymer cis-1,4-polyisoprene, with a molecular weight of 100,000 to 1,000,000 daltons. Typically, a small percentage (up to 5% of dry mass) of other materials, such as proteins, fatty acids, resins, and inorganic materials (salts) are found in natural rubber. Cissus gum has appreciably high emulsion capacity and stability [7]. Acrylic emulsion paint, on the other hand, is a fast-drying paint made of pigments suspended in acrylic polymer emulsion and plasticizers, silicon oils, defoamers, stabilizers, or metal soaps. Acrylic paint, added to wet concrete mix is often used as a cheap way of colouring concrete crafts.

Again, cement, mortal, and concrete-based products are usually brittle. They tend to possess high compressive but relatively low tensile strength. The use of natural fibres for reinforcement is one way of improving the ductility of such products and also promoting sustainable construction [5, 6]. The use of different natural fibres as reinforcement in cement-bonded roofing tiles has been reported by several authors including bamboo by [3], sisal by [15], banana fibre by [12], and coconut husk fibre by [14]. However, another potential source of natural fibre for cement-bonded roofing tile manufacture in Africa and Asia is rattan, a climbing palm with flexible stem generally found near water courses in no less than 20 African countries [2, 13]. Though rattan strands, cane particles and fibres have been used for cement-bonded composite products for interior uses especially for ceiling, floor and wall tile production [1, 10, 11], limited studies have been reported on the rattan fibre-reinforced composite roofing tiles produced with polymer–cement admixtures. In addition to their resistance to fire, these composites have a special attraction for use in warm, humid climates where termites and decay are a major concern.

The aim of this study, therefore, was to investigate the effects of polymer–cement admixtures on selected properties of rattan cane fibre-reinforced roofing tiles.

2 Materials and Methods

Mature rattan (*Laccosperma secundiflorum*) canes (stems) were harvested from the wild forest in south-western Nigeria, and duly identified in the Herbarium of the Department of Botany, University of Ibadan. The canes were cut into 5 cm billets, air-dried to an average moisture content of about 7% and hammer-milled (Plate 1a). The particles obtained were then sieved and particles retained on 2 mm sieve were collected and used. The particles were digested with dilute NaOH (10% w/v) at a temperature of 110^0C and a pressure of one atmosphere for 2 h. The pulp fibres obtained (Plate 1b) were thoroughly washed with potable water and dried in a thermostatic dry box at 60 °C and 85% for 10–15 min to a moisture content of $6.5 \pm 0.1\%$.

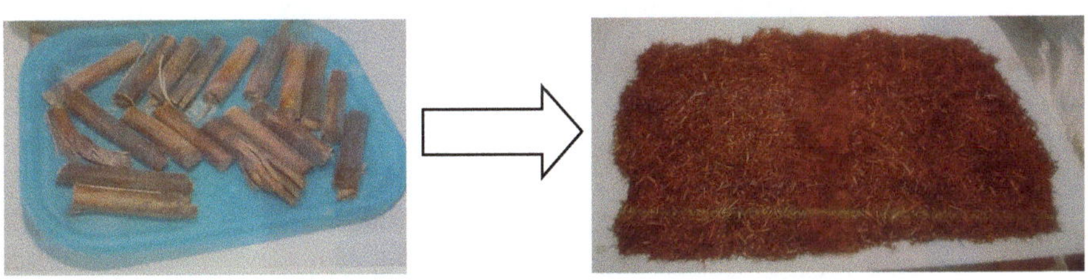

Plate 1 **a** Rattan cane cut into billets, **b** Rattan fibres

Three polymeric materials were used in polymer–cement admixtures—natural rubber latex, *Cissus populnea* gel extracted from the stem of Cissus plant; and acrylic emulsion latex, a synthetic paint. Fresh *Cissus populnea* stems were procured from a local market. They were cut into billets of 20 cm, peeled (Plate 2a) and soaked with portable clean water for three days. After three days, the residues were removed from the viscous adhesive gel *C. populnea*, i.e., the gum (Plate 2b). Commercially sold emulsion and an acrylic were procured and mixed to obtain a solid content to water ratio of 1:0.1 as specified by the manufacturer. Fresh latex (Plate 3) tapped from the rubber tree (*Hevea brasiliensis*) plantation was used. The latex was procured under air tight condition and preserved with Ammonia solution (to prevent coagulation). Fresh Limestone Portland cement packed in 50 kg per bag was purchased from the open market and stored in air tight containers prior to use.

For roofing tile production, dried rattan fibres were manually mixed thoroughly in a container and then mixed with cement at 3% (w/w) mixing ratio. Based on preliminary studies, cement was partially replaced (w/w) with 5, 7.5, 10% natural rubber latex;10, 20%, 30% acrylic emulsion latex; and 10,20,30% *Cissus populnea* gel in different composite samples. De-ionized water was added at 0.5 cement/water ratio. The matrix was spread on a vibrating table using a bricklayer's knife and vibrated at 50 Hz for 60 s to compact the slurry and also get rid of the entrapped air.

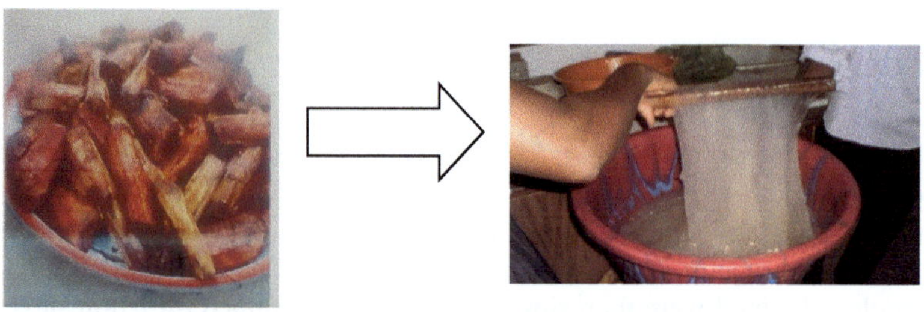

Plate 2 **a** Peeled *C. populnea* billets, **b** *C. populnea* gum

Plate 3 Natural rubber latex

Each cast was covered with a polyvinylchloride (PVC) sheet to reduce voids in the cast. Five replicate samples of 160 mm × 50 mm × 6 mm composite roofing tiles produced were de-moulded after 24 h and cured under wet conditions for 28 days. They were then conditioned at a temperature of 26 °C± 2 °C and a relative humidity of 65% for two weeks as reported by Amoo [4, 11]. The moisture content of each composite was determined using the oven-drying method, while the density was determined as a ratio of mass over volume by measuring the mass (using an electronic precision balance) and volume (using a digital Vernier calliper to measure length, width, and thickness of the samples). For the determination of water absorption and thickness swelling, three dry specimens from each sample were submerged horizontally under 50 mm of distilled water maintained at a room temperature of 26 °C± 2 °C for 24 h, drained for 10 min to remove excess surface water, and then re-weighed as reported by [11]. Water absorption was calculated from the increase in weight of the specimen after submission in water using Eq. 1:

$$\text{Water absorption} = [(W - D)/D] \times 100\% \tag{1}$$

where D = Initial (dry) weight
W = Final (wet) weight

The thickness swelling of each sample was expressed as a percentage of the original thickness using Eq. 2:

$$\text{Thickness swelling} = [(B - A)/A] \times 100\% \tag{2}$$

where A = Original thickness
B = Final thickness

Apparent porosity was determined in accordance with ASTM C-948-81. The dry weight (W_1) of each test sample was measured after which it was transferred into and suspended in a steam vessel for 20 min and weighed (W_2). The steaming was discontinued and the hot water was replaced with cold water (22 ± 2 °C). After 30 min in cold water, each sample was weighed in air (W_3). The apparent porosity was computed using Eq. 3.

$$\text{Pa} = \frac{W3 - W1}{W3 - W2} \times 100\% \tag{3}$$

The moduli of elasticity and rupture of each specimen was determined with the aid of a Universal Testing Machine (UTM) using the 3-point bending test configuration. The 25 KN capacity machine was operated at a cross-head speed of 0.5 mm/min.

Table 1 Effects of cementitious admixtures on density and moisture content

Cementitious admixture	Cement replacement level [%]	Mean composite density [g/cm^3]	Average moisture content [%]
Cement only (control)	0	1.23	11.9 ± 1.2
Acrylic emulsion latex	10	1.08	17.2 ± 1.4
	20	1.30	10.5 ± 0.5
	30	0.98	6.5 ± 0.8
Cissus populnea gel	10	1.08	13.1 ± 1.3
	20	1.10	9.8 ± 0.9
	30	1.53	7.2 ± 1.7
Natural rubber latex	5	0.95	10.8 ± 1.9
	7.5	0.98	10.1 ± 0.7
	10	0.86	10.8 ± 0.6

3 Results and Discussion

3.1 Densities and Moisture Contents of Composite Roofing Tiles

The densities and corresponding moisture contents of the composite roofing tiles are shown in Table 1. The values ranged between 0.86 and 1.23 g/cm^{3}and compared favourably with 0.76–1.34 g/cm^3 reported by [2, 8] for rattan-cement composite boards. Natural rubber latex significantly reduced the density ($p \leq 0.05$), though acrylic emulsion latex (30%) and *C. populnea* gel (10 -20%) also had some density reduction effect. A similar trend was reported by [11]. Hence, as weight reduction is a typical goal in composite roofing tile manufacture, it can be inferred that partial replacement of ordinary Portland cement with 7.5–10% natural rubber latex, 30% acrylic emulsion latex, and 10% *C. populnea* gel is a viable means in achieving this goal. The notable reduction in moisture contents of the roofing tiles containing 20–30% of either acrylic emulsion latex or *C. populnea* gel, was, perhaps, due to their fast-drying nature.

4 Water Absorption and Thickness Swelling

Water absorption is an index for determining the durability of cement-bonded composites since the presence of water can initiate swelling, cracking, and biodegradation of the reinforcing fibre. As shown in Table 2, water absorption values ranged between 0.62 and 1.66%. These values are lower than the upper limit of 25% prescribed by the Indian standard (IS 14276) for cement-bonded composites and 37% recommended

Table 2 Water absorption and thickness swelling of the composite tiles

Cementitious admixture	Cement replacement level [%]	Mean water absorption [%]	Mean thickness swelling [%]
Cement only (control)	0	0.62	0.46 ± 0.01
Acrylic emulsion latex	10	1.48	0.98 ± 0.01
	20	1.38	0.56 ± 0.002
	30	2.66	0.62 ± 0.005
Cissus populnea gel	10	6.12	0.94 ± 0003
	20	4.78	1.52 ± 0.01
	30	7.40	1.70 ± 0.005
Natural rubber latex	5	6.20	1.50 ± 0.04
	7.5	7.70	2.36 ± 0.7
	10	8.40	2.76 ± 0.04

by the Brazilian standard ABNTNBR 7581–1993 for the fibre-reinforced corrugated roofing sheet. An increase in water absorption was observed in all polymer–cement samples compared to the control. However, the water absorption of samples containing the *C. populnea* gel and natural rubber latex was higher the 2% reported in literature for Portland cement-bonded boards [9]. Water absorption is affected by the polarity and size of the material. Lignocellulosic polar materials, which contain OH group in cellulose tend to absorb more moisture. The relatively lower water absorption of the samples containing acrylic emulsion can, therefore, be attributed to the fact that the emulsion must have reduced the amount of hydrophilic OH group in the samples and their capillary porosity. Additionally, emulsion paint usually forms a water-resistant film after drying that might lower water absorption. On the contrary, natural rubber latex and C. *populnea* gum might have entrapped water in the cement matrix of the composites.

Thickness swelling is an important indicator of dimensional stability. The thickness swelling values (Table 2) ranged between 0.46 and 2.76%. Although addition of natural rubber latex above 5% significantly increased the thickness swelling, the values still fell within acceptable limits of 2–3% recommended for cement-bonded composites [8]. The polymers performed in this order: acrylic emulsion latex > *C. populnea* gum > natural rubber latex. Thickness swelling is often directly affected by material properties, e.g., propensity of the material to swell, process variables, e.g., release of compression stress, and interactions between the material and process variables which can modify the swelling behaviour [2, 9]. The generally low swelling rates observed might be due to the visco-elasticity of the polymer matrix. As shown in Fig. 1, there was a direct positive correlation between water absorption and thickness swelling Hence.

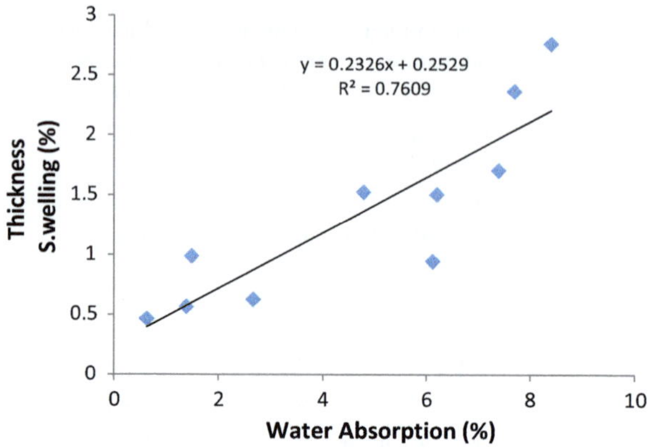

Fig. 1 Correlation between water absorption and thickness swelling of all samples

5 Apparent Porosity

The Apparent porosity values recorded (Table 3) ranged from 11.6 to 28.5%. Except samples containing the *Cissus populnea* gel, apparent porosity of samples produced with polymer–cement admixtures exceeded that of the control. Porosity is a measure of the volume of voids in concrete products and an indication of possible water leakage in roofing tiles. Expectedly, a positive correlation was observed between density and apparent porosity of the *Cissus populnea* gel-based roofing tiles (Fig. 2). What the findings suggest is that except the samples containing the *Cissus populnea* gel, the other composite roof tiles would easily let water through.

Table 3 Apparent porosity of the composite tiles

Cementitious admixture	Cement replacement level [%]	Apparent porosity [%]
Cement only (control)	0	11.6 ± 0.5
Acrylic emulsion latex	10	11.8 ± 0.6
	20	28.5 ± 1.0
	30	15.80 ± 2.5
Cissus populnea gel	10	4.60 ± 0.5
	20	4.10 ± 0.3
	30	7.10 ± 0.3
Natural rubber latex	5	12.20 ± 1.2
	7.5	12.10 ± 1.5
	10	12.60 ± 1.2

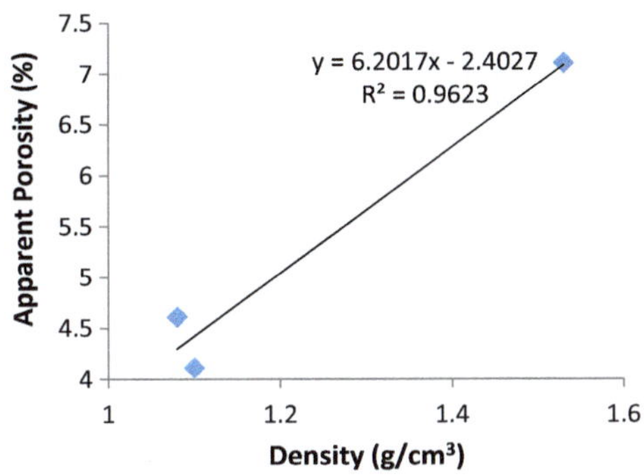

Fig. 2 Correlation between Density and Apparent Porosity of *Cissus populnea* gel composites

6 Moduli of Elasticity and Rupture

The Modulus of Elasticity (MOE) and Modulus of Rupture (MOR) of the composite roofing tiles are presented in Table 4. The MOE values ranged between 930 and 4649 N/mm^2, while the MOR values ranged between 1.12 and 4.58 N/mm^2. These values compare favourably with MOE ($\leq$2078 N/mm^2) and MOR ($\leq$5.2 N/mm^2) reported by [8] for rattan–cement composite boards. Except for composites containing the *Cissus populnea* gel the MOE values obtained were lower than 3000 N/mm^2 recommended in literature for fibre-reinforced composites. The MOR values were also generally lower than 9 N/mm^2 recommended in literature [15]. Except for the *Cissus populnea* gel that increased the MOE and MOR by close to 100% of the control sample values, other polymers significantly lowered the MOE and MOR. Natural rubber latex had the most pronounced negative effects on both composite properties. The observed effects might be attributed to density effects since composite samples containing *Cissus populnea* had the highest density while samples containing rubber latex had the least density. A positive correlation was also observed between MOE and MOR as shown in Fig. 3.

7 Conclusion

Three polymeric materials, i.e., natural rubber latex, *Cissus populnea* gel, and acrylic emulsion latex were tested as a partial replacement for Portland cement in rattan fibre-reinforced roofing tile production. Selected sorption and mechanical properties of the composite roofing tiles were determined. Results obtained show that relatively strong and dimensionally stable rattan fibre-reinforced roofing tiles can be produced with the *Cissus populnea* gel–cement admixtures at cement replacement levels of up to 30%.

Table 4 MOE and MOR of the composite tiles

Cementitious admixture	Cement replacement level (%)	Mean modulus of elasticity [N/mm^2]	Mean modulus of rupture [N/mm^2]
Cement only (control)	0	2054	2.31 ± 0.02
Acrylic emulsion latex	10	1492	1.90 ± 0.03
	20	1665	1.37 ± 0.07
	30	808	1.27 ± 0.05
Cissus populnea gel	10	4392	3.94 ± 0.5
	20	4649	3.30 ± 0.4
	30	4232	4.58 ± 0.2
Natural rubber latex	5	930	1.12 ± 0.05
	7.5	1373	1.61 ± 0.04
	10	1658	1.70 ± 0.08

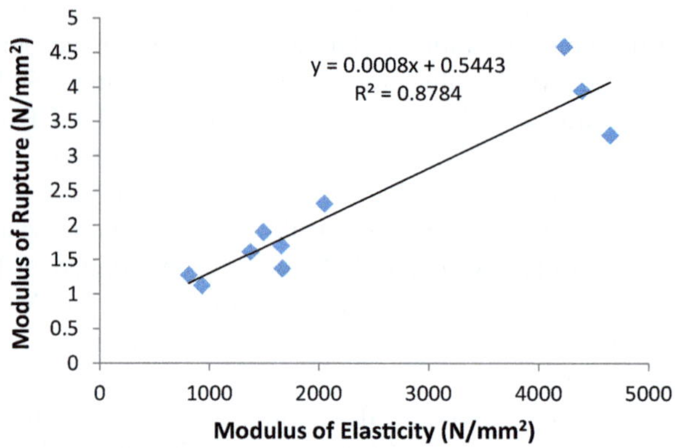

Fig. 3 Correlation between MOE and MOR of all Samples

References

1. Adefisan OO, Olorunnisola AO (2007) Compatibility of two rattan species (*Laccosperma secundiflorum* and *Calamus derratus*) with portland cement for particleboard production. J Appl Sci Eng Technol 7(1):50–56
2. Adefisan OO, Olorunnisola AO (2012) Strength and sorption properties of cement bonded composites made from Calamus deeratus and Laccosperma secundiflorum Canes. ASUU J Sci 1(1):51–66
3. Alade GA, Olutoge FA (2004) Bamboo fibre-reinforce cement used as a roofing sheet. J Civil Eng Res Pract 1(2):107–118
4. Amoo K, Adefisan OO, Olorunnisola AO (2016) Development and evaluation of cement-bonded composite tiles reinforced with Cissus populnea Fibres. Int J Compos Mater 6(4):133–139. https://doi.org/10.5923/j.cmaterials.20160604.06
5. Balaguru PN, Shah SP (1992) Fiber reinforced cement composites. McGraw Hill, New York

6. Bentur AO, Aekers SA (1989) The microstructure and ageing of cellulose fibre -reinforced cement composites cured in a normal environment. Int J Cement Compos Lightweight Concrete 11(2):99–110
7. Iwe MO, Obaje PO, Akpapunam MA (2004) Physicochemical properties of cissus gum powder extracted with the aid of edible starches. Plant Foods Hum Nutr 59:161–168
8. Olorunnisola AO, Pitman A, Mansfield-William H (2005) Strength properties and potential uses of rattan-cement composites. J Bamboo Rattan 4(4):343–352
9. Olorunnisola AO (2006) Strength and water absorption characteristics of cement-bonded particleboard produced from coconut husk. J Civil Eng Res Pract 3(1):41–49
10. Olorunnisola AO, Agrawal SP (2015) Effects of NaOH concentration and Fibre content on Physico-mechanical properties of cement-bonded rattan fibre composites. Pro Ligno Int J Wood Eng 11(4):192–198
11. Olorunnisola AO, Asimiyu AK (2016) Effects of Cissus populnea Gum and rubber latex on Physico-mechanical properties of cement-bonded rattan composites. J Civil Eng Constr Technol 7(3):20–27. https://doi.org/10.5897/JCECT2015.0395
12. Olorunnisola AO, Ope-Ogunseitan F (2016) Effects of lime and $CaCl_2$ on impact strength and dimensional stability of banana fibre- reinforced composite roofing tiles. Proceedings of the 15th Inorganic-Bonded Fiber Composites Conference (IIBCC) held in at Fuzhou Empark Exhibition Grand Hotel, Fuzhou, China, November 8–11, 2016. pp 76–84
13. Olorunnisola AO (2019) Development of sustainable building materials from agro-industrial wastes in Nigeria. In: Hemeda S (ed) sustainable construction and building materials Published online and in hard copy by INTECHOPEN. https://doi.org/10.5772/intechopen.78713, ISBN: 978–1–78985–750–4, Print ISBN: 978–1–78985–749–8. https://www.intechopen.com/books/sustainable-construction-and-building-materials, pp 55–74
14. Olorunnisola AO, Adeniji AO (2019) investigations on the effects of cement replacement and calcium chloride addition on selected properties of coconut husk fibre-reinforced roofing tiles. Mater Res Proc 11:253–259. https://doi.org/10.21741/9781644900178-21
15. Savastano Jr H, Warden PG, Coutts RSP (2003) potential of alternative fibre cements as building materials for developing areas. Cement Concrete Compos 25:585–592

Effects of Selected Cement Admixtures and Accelerated Curing on Physico-Mechanical Properties of Coconut Husk Fibre-Reinforced Composite Roofing Tiles

Anthony O. Adeniji, Abel O. Olorunnisola, and Holmer Savastano

Abstract The ban on asbestos-cement roofing sheets in many countries has contributed to the search for more environmentally friendly roofing materials. This study examined the effects of different cement admixtures and accelerated curing on physico-mechanical properties of coconut husk fibre-reinforced composite roofing tiles. Incineration temperatures were 800 °C for Rice husk Ash (RHA), 500 °C for Chicken Eggshell Ash Type 1(CESA1) and 900 °C for Chicken Eggshell Ash Type 2 (CESA2) which was hydrolysed before use. Calcium Carbide Waste (CCW) and Limestone Portland Cement (LPC) were also used as admixtures. For composite manufacture, fibre content was 4% of cement mass; cement: water mass ratio was 0.4, and cement: sand ratio was 1:2. A super-plasticizer of 0.3% of cement mass was added. Cement admixtures were 70% OPC + 15%RHA + 15%CESA; 70%LPC + 15%CCW + 15%CESA; and 70%LPC + 7.5%RHA + 7.5% CCW + 15% CESA. Control specimens were thermally cured at 60 °C for 5 days. The specific gravity and Thermal-Degradation-Temperature (TDT) of raw materials were determined. The CO_2- cured specimens were exposed to 15% CO_2 at 60 °C, 60% RH and 0.34 MPa for 9 h. Six, 160 × 40 × 6 mm^3 replicate samples were tested to determine composite Bulk Density (BD), Moisture Contents (MC), Water Absorption (WA), and Modulus of Rupture (MOR). Micrographs of specimens were obtained using Scanning Electron Microscope (SEM). Composite bulk density (1.77–2.14 g/cm^3) and WA (6.9–16.1%) were relatively high. Partial replacement of cement with RHA, CCW and CESA reduced the bulk density and increased the WA, particularly in the thermally cured CESA1-based samples. The MOR (9.2–12.9 MPa) of the CO_2-cured CESA2 samples were higher than the CESA1 samples and control. Micrographs showed acceptable fibre–matrix interaction in all tested samples.

A. O. Adeniji (✉) · A. O. Olorunnisola
Department of Wood Products Engineering, University of Ibadan, Ibadan, Nigeria
e-mail: toneedenk@gmail.com

H. Savastano
Faculty of Biosystem Engineering, University of Sao Paulo, Pirassununga, Brazil
e-mail: holmersj@usp.br

M. Jawaid et al. (eds.), *Proceedings of 2nd World Conference on Byproducts of Palms and Their Applications*, Springer Proceedings in Materials 19,
https://doi.org/10.1007/978-981-19-6195-3_11

Keywords Coconut husk · Rice husk ash · Chicken eggshell ash · Carbide waste · Roofing tiles

1 Introduction

Providing housing for the teeming population is one of the major challenges in many countries across Africa. The relatively high cost of conventional building materials is a factor affecting housing provision on the continent. Roofing constitutes a major cost component in building construction. While asbestos sheets have been outlawed in many countries because of their carcinogenic characteristics, stone-coated aluminium tiles, long-span aluminium sheets and other conventional roofing materials are relatively expensive and thus less affordable for a majority of low-income earners in Nigeria. One way of addressing this challenge is to develop alternative roofing materials from Cement-Bonded Composites (CBCs). CBCs can be produced with natural fibres used for reinforcement and industrial waste products used as pozzolans for a partial replacement for cement. Some of the objectives of incorporating fibrous materials i.e. coconut husk fibre in the matrix of CBCs are to increase the toughness, reduce thermal conductivity, and reduce the cost of production as well as the weight of the tiles [1–4]. Reported studies on the durability of natural fibre composites using mechanical strength and microstructural properties also concluded that natural fibres enhance the strength and flexural performance of the concrete [5].

Coconut (*Cocos nucifera* L.) is commonly called the tree of life because of its myriad uses. It grows to a height of 20 m and takes about 12 months for the fruit to develop from flowering to maturity [6]. The average mature coconut weighs 680 g and about 42% of its weight is made up of the husk [7]. It has been reported in the literature that coconut husk fibre retains a higher percentage of its initial strength properties than all other fibres after long exposure to various media [8]. Significant increase in toughness has also been reported for composites made with the incorporation of short coconut fibre and ordinary Portland cement matrix [9].

Different materials have been tested and found suitable as pozzolans or admixture materials for partial replacement of Limestone Portland cement (LPC) in concrete works over the years. Examples include: Rice Husk Ash (RHA), Calcium Carbide Waste (CCW), and Eggshell Ash (ESA). RHA contains 85–90% of silica which is mostly amorphous [10]. Calcium carbide waste is mainly composed of calcium hydroxide with a mass fraction above 92% and is highly alkaline (pH > 12), while eggshell contains 95% of calcium carbonate in the form of calcite [11]. The RHA could be used as a micro-filler in concrete mixtures and noted that partial replacement of the cement with the relatively lighter RHA resulted in a decrease in concrete density and permeability [12]. It was also reported that rice husk and eggshell ash when used individually as partial replacements for cement produced concrete whose strength properties were somewhat equal to that of conventional concrete [13]. Mixing calcium carbide waste with cement could yield pozzolanic reactions resulting in final products that are similar to those obtained from the cement hydration process [14]. It was reported [15, 16] that the incorporation of 5% eggshell powder in concrete production resulted in an optimum concrete strength. Incineration of the eggshell at about 900 °C produces calcium oxide which can be used as an admixture in concrete. The eggshell ash when used as an admixture with cement exhibited the properties of a good accelerator because of the extra calcium oxide provided by the ash [17].

The aim of this study was to study the effects of CO_2 curing and partial replacement of cement on the selected properties of coconut husk fibre-reinforced composite roof tiles.

2 Methodology

Coconut fibres were removed from the husk and separated into individual strands which were cut into an average length of about 19 mm. Rice husk was oven dried for 24 h at 60 °C and incinerated at 800 °C into white ash. Carbide waste obtained from a mechanical workshop was air-dried, pulverized, and sieved with a 75 μm sieve. Eggshells collected from a restaurant were oven-dried at 60 °C for 24 h and incinerated at 500 °C and 900 °C, respectively, for ash production. For eggshell ash produced at 900 °C, water at a ratio of 1 part to 3 parts of calcium oxide based on molar mass was added to hydrolyse the eggshell ash. Heat loss was observed in the beaker and it was allowed to cool for about 10 min before using it for composites production. For the control (100%) cement and experimental samples, coconut fibre of 4% (mass of cement) was mixed with Portland cement. Cement–water ratio was fixed at 1:0.4 while the cement–sand ratio was fixed at 1:2. For the production of control samples, 100% cement was mixed with coconut fibre, sand, and water. Additionally, for the experimental samples, cement was partially replaced with an admixture of 15% rice husk ash +15% eggshell ash, 15%calcium carbide waste +15% eggshell ash, and 7.5% rice husk ash +7.5%calcium carbide waste +15% eggshell ash (by mass of cement), respectively. Meflux of 0.3% by mass of cement was added to water for mixing the composite matrix.

For composite manufacture, each mixture was poured into a $6 \times 160 \times 300$ mm^3 mould and spread uniformly to a thickness of 6 mm, vibrated for 60 s to reduce the porosity in the composite and smoothened with a hand trowel. The composite was then de-moulded and cured under saturated water vapour at 25 ± 2 °C (in a sealed plastic bag) for 48 h (this was achieved by spraying the composite in a plastic bag with water before sealing). A set of samples and the control were subjected to thermal curing while another set of samples was subjected to accelerated carbonation. For thermally cured samples, composites enclosed in a sealed plastic bag were placed inside a chamber maintained at a temperature of 60 °C for five days. Composite samples subjected to accelerated carbonation were placed in a climatic chamber at a temperature of 60 °C and relative humidity of 60% for 9 h The CO_2 content in the climatic chamber was regulated to about 15% volume at 0.34 MPa. Phenolphthalein solution was prepared at 2%, and diluted in anhydrous ethanol [18] to determine the degree of penetration of CO_2 in the composite by a piece of broken sample and sprayed with phenolphthalein. The composite samples were removed from the plastic bag and cut into 6 × 40 × 160 mm^3 samples in six replicates. Physical properties and thermal degradation behaviour of raw materials were tested using thermogravimetric analysis [19]. The samples were tested for bulk density, moisture contents, water absorption, and modulus of rupture using standard methods [20–22] (ASTM D 1037 (1991), ASTM D 1037/89 (1991), ASTM D3043 00(2011)). Scanning Electron Microscope (SEM) was used for microstructural investigation of the fibres by observing the interactions between the fibre and matrices mixtures. Data were analysed using descriptive and inferential statistics, and ANOVA at $\alpha_{0.05}$.

3 Results and Discussion

3.1 Specific Gravity and Moisture Contents of Raw Materials

As shown in Table 2, the specific gravity of the coconut fibre at 8.2% moisture content was 1.38. On the other hand, the specific gravity of calcium carbide at 1.9% moisture content was 2.29, which is comparable with the specific gravity of 2.3 [23]. Also, the specific gravity of the cement was 3.1 at 1.3% moisture content which is comparable to the value of 3.2 [24]. Lastly, the specific gravity of rice husk ash was 2.2 at 1.4% moisture content which is lower than that of calcium carbide waste and cement and suggests that admixtures in which cement is partially replaced with rice husk ash would be lighter than those containing calcium carbide waste (Table 1).

3.2 Thermal Properties of Raw Materials

Figures 1, 2, 3 and 4 show the Thermal-Degradation-Temperature (TDT) curves for coconut fibre, rice husk ash, calcium carbide, and chicken eggshell ash. There was a moisture loss of about 0.7% in the coconut husk fibre between 20 °C and 80 °C. The second weight loss was observed between 110 °C and 500 °C while maximum weight loss of about 41.0% was observed between 760 °C and 900 °C. These results are comparable with the thermal behaviour of wood pulp fibre [25]. It has also been reported that the initial decomposition of the cellulose components takes place mostly in the amorphous region which is similar to the result of coconut fibre [26]. For rice husk ash, moisture loss was observed between 25 °C and 65 °C followed by maximum weight loss of about 19.0% at temperatures between 100 °C and 880 °C. The thermal degradation of rice husk ash might be a result of larger particle sizes which yield more residue because of poor heat transfer [27, 28]. The calcium carbide degradation curve showed moisture loss of 2.0% between 20 °C and

Table 1 Percentage formulations for samples composite

Samples	Formulations
Samples containing eggshell ash obtained at 500 °C and thermally cured at 60 °C	
A	100% PLC
B	15%RHA + 15% CESA + 70% PLC
C	15% CCW + 15% CESA + 70% PLC
D	7.5% RHA + 7.5% CCW + 15% ESA + 70% PLC
Samples containing eggshell ash obtained at 500 °C and subjected to accelerated carbonation	
A1	100% PLC
B2	15% RHA + 15% CESA + 70% PLC
C1	15% CCW + 15% CESA + 70% PLC
D1	7.5 RHA + 7.5% CCW + 15% ESA + 70% PLC
Samples containing eggshell ash obtained at 900 °C and subjected to accelerated carbonation	
A2	100% PLC
B2	15% RHA + 15% CESA + 70% PLC
C2	15% CCW + 15% CESA + 70% PLC
D2	7.5%RHA + 7.5% CCW + 15% CESA + 70% PLC

*Water/cement was expressed in ratio, others in percentage
*RHA = Rice husk ash; CESA = Chicken eggshell; CCW = Calcium carbide Waste; PLC = Portland Limestone Cement

Table 2 Specific gravity of coconut fibre, calcium carbide, cement, rice husk ash

Raw material	Specific gravity	Moisture contents [%]
Coconut fibre	1.38	8.2
Calcium carbide	2.29	1.9
Cement	3.10	1.3
Rice husk ash	2.20	1.4
Eggshell ash	2.49	2.4

120 °C, followed by weight loss between 150 °C and 400 °C. Additional weight loss was observed between 490 °C and 740 °C of about 6.0%. The eggshell ash showed moisture loss at the initial stage of 5.2% between 20 °C and 100 °C followed by weight loss of 5% between 200 °C and 400 °C. Further degradation was observed between 440 °C and 650 °C resulting in a weight loss of 5.0%. This trend of results is similar to the thermal degradation of Quail eggshell powder [29].

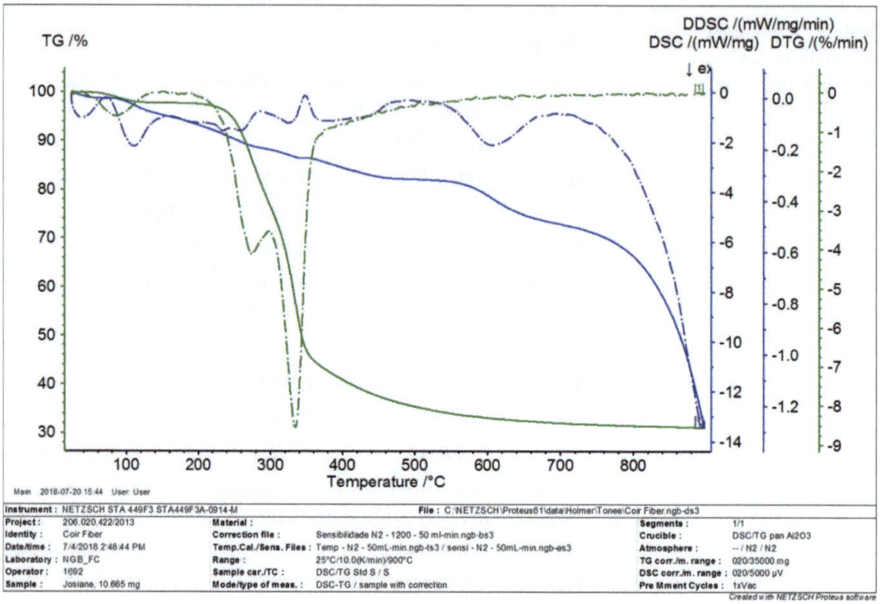

Fig. 1 TDT of coconut fibre

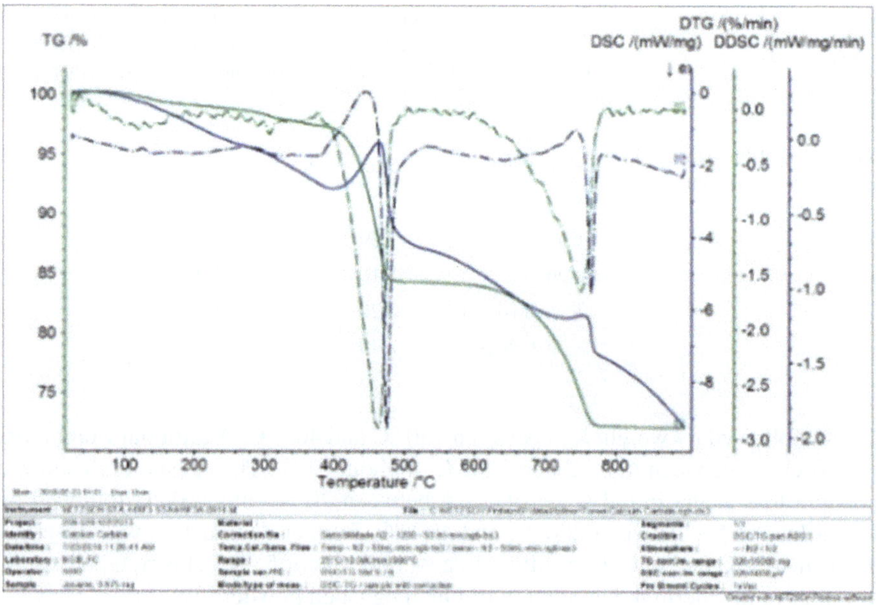

Fig. 2 TDT of calcium carbide

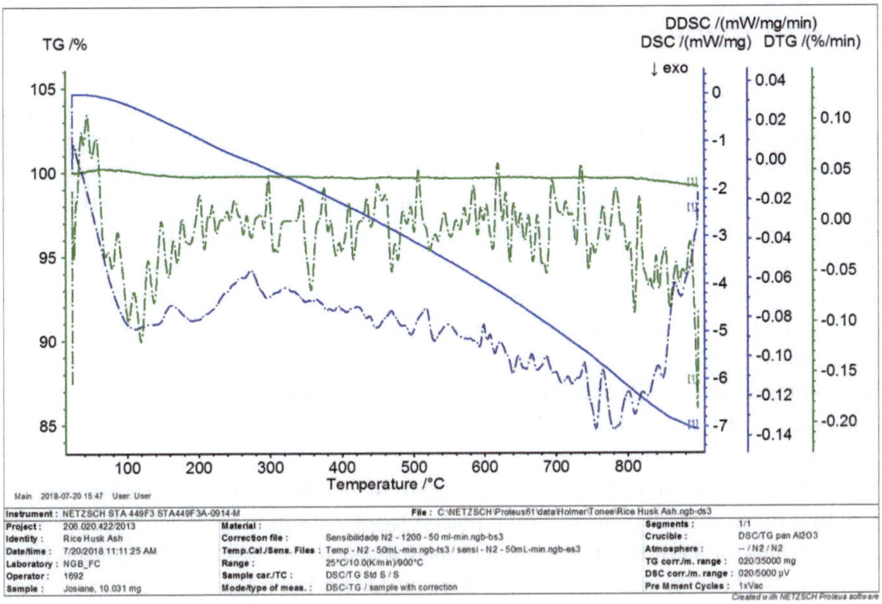

Fig. 3 TDT of rice husk ash

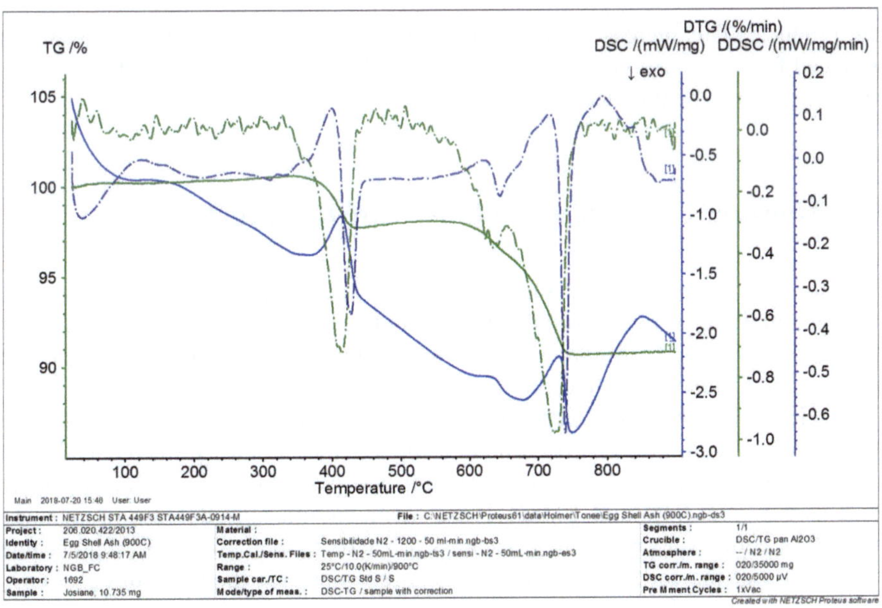

Fig. 4 TDT of eggshell ash

3.3 Bulk Density of Composite Roofing Tiles

The average bulk densities for composites subjected to different curing conditions are presented in Table 1. The bulk densities of the thermally cured samples ranged between 1.77 gcm^{-3} and 1.94 gcm^{-3}. Expectedly, the sample containing rice husk ash had the lowest bulk density, while the control had the highest bulk density. The bulk densities of samples in which the cement was partially replaced were compared favourably with values for composites reinforced with green coconut fibre (1.7–1.9 cm^{-3}) [30] and were slightly higher than 1.38 $gcm^{-3}$for composites reinforced with Sisal (*Agave sisalana*) [31]. Samples containing rice husk ash had the same bulk density (1.8 gcm^{-3}) as composites reinforced with *Eucalyptus grandis* [30]. The observed lower bulk density for composites containing calcium carbide waste is in agreement that calcium carbide waste addition to the composite resulted in a lower bulk density [32].

Accelerated carbonation generally increased the bulk densities of the samples including the control whose value increased from 2.03 gcm^{-3} to 2.14 gcm^{-3}. The bulk densities of samples in which cement was partially replaced by eggshell ash incinerated at 500 °C and subjected to accelerated carbonation were generally high while samples containing rice husk ash had the lowest bulk density. The observed increase in bulk densities might be a result of carbonation which is in tandem [33] that carbonation reactions promote the filling of the pores in the matrix with carbonated products. This increases the bulk density because calcium carbonate ($CaCO_3$) produced from the carbonation is denser than calcium hydroxide ($Ca(OH)_2$) that is released in the cement hydration.

3.4 Moisture Contents of the Composite Roofing Tiles

The moisture contents of the samples subjected to different curing regimes are presented in Table 3. For thermally cured samples, the moisture content range was between 9.5 and 16.1%. Samples containing rice husk ash had the highest moisture contents attributable to the hydrophilic nature of the rice husk ash [34].

Accelerated carbonation generally reduced the moisture contents of all the composites including the control. The moisture content (control) was reduced from 9.5% to 6.8%. Composites in which cement was partially replaced had moisture contents between 7.4% and 10.9%. There was a significant reduction in moisture contents of composites containing rice husk ash and eggshell ash. This might be due to the fact that the carbonation process required water for the dissolution of CO_2 into the pores by diffusion of CO_2 into the empty pores which dissolved in the pore solution and then reacted with OH^- to form $CO_3{}^{2-}$ thereby resulting in moisture loss [35]. The standard [36] specified that the moisture contents of fibre cement composite must be lower than 12%. All the composites subjected to accelerated carbonation

Table 3 Bulk density, moisture contents, water absorption and modulus of rupture (MOR) of composite tiles production under different curing methods

Formulation	Bulk density [gcm^{-3}]	Moisture contents [%]	Water absorption [%]	MOR [MPa]
Samples containing eggshell ash obtained at 500 °C and thermally cured at 60 °C				
A	2.03 ± 0.17^{B}	9.5 ± 0.17^{A}	9.4 ± 0.17^{A}	7.4 ± 0.48^{A}
B	1.77 ± 0.03^{A}	16.1 ± 1.08^{B}	16.1 ± 1.08^{C}	1.7 ± 0.44^{C}
C	1.94 ± 0.01^{B}	$11.3 \pm 0.33C$	11.3 ± 0.33^{B}	5.1 ± 0.37^{B}
D	1.87 ± 0.02^{B}	$12.8 \pm 0.98^{D}t$	12.8 ± 0.98^{B}	2.1 ± 0.37^{B}
Samples containing eggshell ash obtained at 500 °C and subjected to accelerated carbonation				
A1	$2.14 \pm 0.02^{A, B}$	6.8 ± 0.10^{A}	6.9 ± 0.09^{A}	10.4 ± 2.20^{A}
B1	1.93 ± 0.01^{A}	10.9 ± 0.31^{B}	10.9 ± 0.32^{C}	8.3 ± 1.82^{AB}
C1	2.06 ± 0.01^{B}	7.4 ± 0.11^{A}	7.4 ± 0.11^{A}	9.4 ± 2.1^{AB}
D1	1.97 ± 0.05^{B}	9.7 ± 1.05^{C}	9.7 ± 1.05^{B}	7.6 ± 1.75^{B}
Samples containing eggshell ash obtained at 900 °C and subjected to accelerated carbonation				
A2	$2.13 \pm 0.01^{A, B}$	7.3 ± 0.13^{A}	7.3 ± 0.14^{A}	9.2 ± 1.7
B2	2.02 ± 0.01^{A}	8.7 ± 0.34^{B}	8.7 ± 1.10^{C}	11.2 ± 1.10^{B}
C2	$2.07 \pm 0.01^{A, B}$	8.1 ± 0.33^{A}	8.1 ± 0.33^{B}	12.9 ± 1.34^{C}
D2	2.04 ± 0.01^{AB}	8.4 ± 0.38^{B}	8.5 ± 0.33^{C}	10.0 ± 1.72

Mean with the same letters are not statistically different

for eggshell incinerated at 500 °C and 900 °C had their values lower than the specified percentage. A positive correlation was observed between the bulk density and moisture contents of composite tiles as shown in Fig. 5.

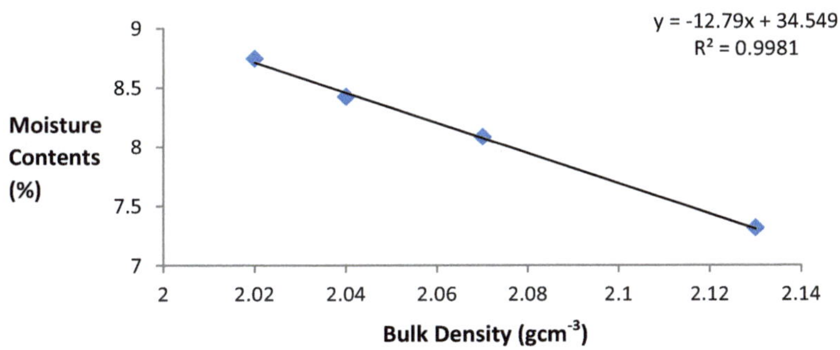

Fig. 5 Correlation between bulk density and moisture contents for composite tiles

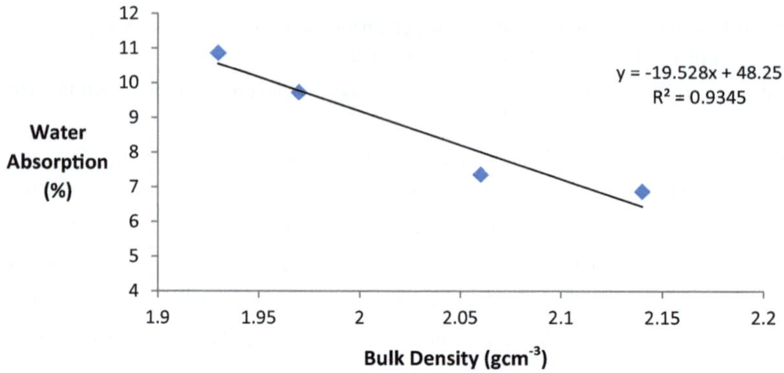

Fig. 6 Correlation between bulk density and water absorption for composite samples subjected to accelerated carbonation

3.5 *Water Absorption of Composite Roofing Tiles*

Table 3 shows the water absorption of the composite samples cured under different regimes. For the thermally cured samples, the range of water absorption was between 9.4% and 16.1%. There was a positive correlation between bulk density and water absorption for all the composites (Fig. 5). The range of water absorption values is in tandem with what was reported in the literature for water absorption (4.6–16%) of composites reinforced with bagasse [37–39] but less than water absorption (21.2–30.5%) for composites reinforced with green coconut fibre [30]. Composites in which the cement was partially replaced with calcium carbide waste and eggshell ash had the lowest water absorption. The higher water absorption of rice husk ash samples might be due to their low bulk density and higher porosity [40, 41].

Accelerated carbonation generally reduced the water absorption except for the control samples. The range of water absorption values for carbonated samples was lower than 23.3–33.3% reported in the literature for composites reinforced with sisal [42] and also lower than 30.9–36.6% was reported for composites produced from *Eucalyptus grandis* [43]. The composites containing rice husk ash had a higher water absorption value than samples containing calcium carbide perhaps due to the silica oxide content of rice husk ash being higher than that of calcium carbide [44]. The water absorption values obtained in this study were generally lower than 37% as recommended [45] for fibre-reinforced corrugated roofing sheets. It can be concluded that composite samples could be used for exterior purposes (Fig. 6).

3.6 *Modulus of Rupture*

The moduli of rupture (MOR) of the composites are presented in Table 3. For thermally cured samples, MOR ranged between 1.7 MPa and 7.4 MPa. The values were

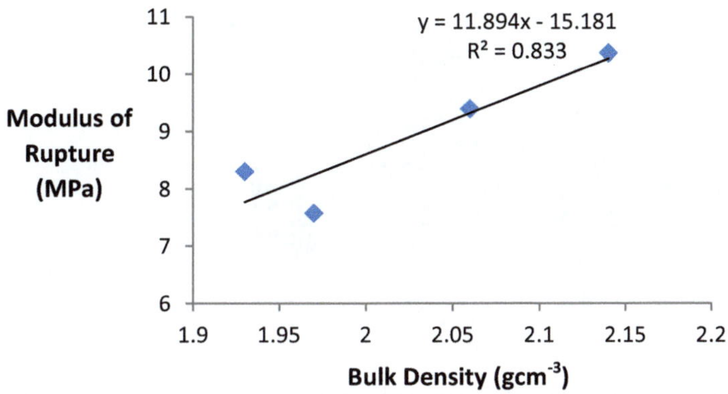

Fig. 7 Correlation between bulk density and modulus of rupture for composite samples

lower than 11.0 MPa for composites reinforced with coconut fibre and cellulosic pulp [30]. Samples containing calcium carbide waste had relatively higher MOR values than samples containing RHA, carbide waste, and eggshell ash. The MORs of samples containing calcium carbide waste and eggshell ash compared favourably with a composite from sisal kraft (13.1 MPa) [43]. For all the composite samples subjected to thermal curing, the MORs were lower than the 7.7 MPa reported for asbestos–cement ceiling board [46] but compared favourably with 0.8–7.4 MPa reported [3] for composites reinforced with rattan (*Lacoperma secundiflorum*) particles.

Accelerated carbonation resulted in improved MOR in all the composites, including the control. This might be because carbonation improved the contact between fibres and cement matrix and favoured better adhesion [30]. There was a considerable increase in the MOR of samples containing rice husk ash and eggshell ash. An increase in the incineration temperature of eggshell ash to 900 °C also enhanced the MOR. This might have been a result of the effects of extra calcium oxide produced from eggshell (calcium carbonate) incinerated at 900 °C which acted as a good accelerator and improved the setting time [47]. There was a significant difference in MORs between composite tiles cured under accelerated carbonation. Figure 7 showed a positive correlation between the bulk density and MOR.

3.7 *Microstructural Analysis*

Microstructural images of coconut fibres in the matrix are shown in Fig. 8. Figure 8 shows that samples containing calcium carbide and eggshell ash and subjected to accelerated carbonation had lesser pores around the coconut fibre root. This might have accounted for their higher MOR (12.9 MPa) and suggests the existence of strong fibre–matrix interaction in the composite materials. On the contrary, samples containing rice husk ash and eggshell ash had more pores around the fibre root

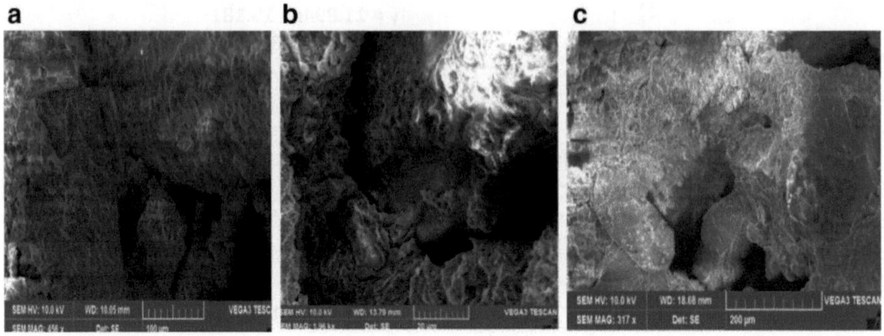

Fig. 8 **a–c** show the SEM images for composites containing calcium carbide–eggshell ash, rice husk ash–eggshell ash and calcium carbide waste–rice husk ash–eggshell ash

which might have accounted for their relatively lower MOR (11.2 MPa). Fibre–matrix interaction observed in the composite samples containing calcium carbide and eggshell ash might be responsible for lower water absorption compared with other experimental samples.

4 Conclusion

This study showed that:

1. Cementitious admixtures containing rice husk ash, calcium carbide waste, and chicken eggshell ash are suitable for coconut fibre-reinforced composite roofing tile production.
2. The bulk density of the composite roofing tiles might be reduced when RHA, CCW and CESA are used as partial replacements for cement.
3. Carbonation of coconut fibre-reinforced composite roofing tiles enhances the moduli of rupture and reduction in water absorption which in turn improves the quality of roofing tiles.

References

1. Savatsano H Jr, Agopyan V (1999) Transition zone studies of vegetable fibre-cement paste composites. CemConcr Compos. 21(1):49–57
2. Khedari J, Suttisonk B, Charoenvai S (2000) New lightweight composite construction materials with low thermal conductivity. CECO J 804:1–6
3. Olorunnisola AO (2005) Dimensional stability of cement-bonded composite boards made from rattan cane particles. J Bamboo Rattan 4(2):173–185
4. Olorunnisola AO (2006) Strength and Water absorption characteristics of cement-bonded particle board produced from coconut husk. J Civil Eng Res Pract 3(1):41–49

5. Sivaraja M, Kandasamy S (2009) Characterisation of natural fibres as concrete composites for structural applications 36(1–4):385–395
6. Agustin YV (2004) United Coconut Association of the Phillipines (UCAP) and Covers www.ucap.org.ph
7. Badmus GA, Adeyemi NA, Owolarafe OK (2007) Development of an improved manual dehusking lever. Nigeria J Palms Seeds 16:36–47
8. Ramakrishna G, Sundararajan T (2005) Studies on the durability of natural fibres and the effect of corroded fibres on the strength of mortar. Cement Concr Compos 27(5):575–582
9. Abdulahi S, Baharin J, Mazlee MN, Kamarudin H (2011) Composite cement reinforced coconut fibre: physical and mechanical properties and fracture behaviour. Aust J Basic Appl Sci 5(7):1228–1240
10. Narayan PS (2002) Use of Rice Husks for Concrete moulding. Res Indus 20(3):142–150
11. Arias L, Quijada R, Yazomdani-Pedram M, Toro P (2006) Patent Appl. US. Universidad de Chile. Invs J, 2006–068185-A1
12. Maisarah A, Mohd SF, Siti AS (2015) Effect of Rice Husk Ash (RHA) on physical property and mechanical strength of concrete. Adv Mater Res 1115:150–155
13. Jayasankar R, Mahindran N, Ilangovan R (2010) Studies on concrete using fly ash, rice husk ash and eggshell powder. Int J Civil Struct Eng Integr Pub Serv. ISSN 0976–4399, 1(3):362–372
14. Wang YL, Dong SJ, Liu LL, Cui SP (2013) Using calcium carbide slag as one of calcium-containing raw materials to produce cement clinker. Energy Environ Mater 171:743–744
15. Amarnath Y (2014) Properties of concrete with eggshell powder as cement replacement. Indian Concrete J
16. Gowsika D, Sarankokila S, Sargunan K (2014) Experimental investigation of eggshell powder as a partial replacement with cement in concrete. Int J Eng Trends Technol (IJETC) 14(2)
17. Mtallib MOA, Rabiu A (2009) Effect of eggshell ash on the setting time of the cement. Nigeria J Technol 28(2):29–38
18. Agopyan V, Savastano (Jr) H, John VM, Cincotto MA (2005) Development on vegetable fibre-cement based materials in Sao-Paulo, Brazil: an overview. Cement Concrete Compos 27(5):527–536
19. ASTM E-2550–17 (2017) Standard Test Method for thermal stability by thermogravimetry, ASTM International, West Conshohocken, PA
20. ASTM D1037–12 (2012) Standard test methods for evaluating properties of wood-base fibre and particle panel materials, USA
21. ASTM D1037–99 (1999) standard test method for evaluating properties of wood-base fibre and particle panel material, USA
22. ASTM D3043–00 (2000) Standard test method for structural panels in flexure
23. Makaratat N, Jaturapitakkul C, Namarak C, Sata V (2011) Effects of binder and $CaCl_2$ contents on the calcium carbide residue—fly ash. Cement Compos 33:436–449
24. Charoenvai S, Khedari J, Hirunlabh J, Daguenet M, Quenard D (2005) Impact of rice husk ash on durian fibre-based construction materials IUDBMC. International conference on durability of building materials and components. LYON (France), 17–20th April (2005)
25. Stevulova N, Hospodarova V, Estokova A (2016) Study of thermal analysis of selected cellulose fibres 3:18–21
26. Jonoobi M, Niska KO, Harun J, Mistra M (2009) Chemical composition, crystallinity and thermal degradation of bleached and unbleached kenaf bast (*Hisbiscus cannabinus*) pulp and nano fibres. Bio-Resources 4(2):626–639
27. Stenseng M, Jensen A, Dam-Johansen K (2001) Investigation of biomass pyrolysis by thermo-gravimetric analysis and differential scanning calorimetry. J Anal Appl Pyrol 58–59:765–780
28. Stresov V, Moghtaderi B, Lucas J (2003) Thermal study of decomposition of selected biomass samples. J Therm Anal Calorim 72:1041–1048
29. Leandro MC, Juan AC, Enrique R, Celio LC, Rodrigo SV (2017) Relevance of physicochemical properties of calcined quail eggshell (CaO) as a catalyst for biodiesel production. J Chemi 12
30. Pereira CL, Savastano (Jr) H, Paya J, Santos SF, Borrachero MV, Monzo J, Soraino L (2013) Use of highly reactive rice husk ash in the production of cement matrix reinforced with green coconut fibre. Indus Crops Products 49:88–96

31. Roma LC, Martello (Jr) LS, Savastano (Jr) H (2008) Evaluation of mechanical, physical and thermal performance of cement-based tiles reinforced with vegetable fibre. Constr Build Mater 22:668–674
32. Olorunnisola AO, Ogundipe AI (2012) Impact and water resistance of rattan composites produced with rice husk ash and welder's used waste carbide as partial replacement for Portland cement. A paper presented at (ECOWOOD) 2012 -5th international conference on environmentally—ompatible forest products. Held at Fernando Pessog, University, Porto, Portugal pp 199–209
33. Almeida AEFS, Tonoli GHD, Santos SF, Savastano (Jr) H (2013) Improved durability of vegetable fiber reinforced cement composite subject to accelerated carbonation at early age. Cem Concr Compos 42:49–58
34. Aznizam A, Azman H, Ahmad-Fuad M (2005) Effects of accelerated weathering on the mechanical properties of oil palm empty fruit bunch filled UPVC composite. Iran Polym J 14(4):627–635
35. Frias M, Gonis S (2013) Accelerated carbonation effect on behavior of tertiary Portland cement. Compos B 48:122–128
36. ISO 8335–87 (1987) Cement-bonded particleboards-board of Portland or equivalent cement reinforced with fibrous wood particles
37. Moslemi AA, Souza M, Geimer R (1994) Accelerated ageing of cement-bonded particle board. 4th international inorganic-bonded wood-fibre composite materials conference, spokane, Washington, USA, pp 83–83
38. Ajayi B (2002) Preliminary investigation of cement-bonded particle board from maize stalk residues. Nigeria J Forestry 32(1):33–37
39. Ajayi B (2003) Investigation of the dimensional stability of cement-bonded composite board fabricated from coffee husks. Nigeria J Forestry 33:88–93
40. Olorunnisola AO, Agrawal S (2015) Effects of NaOH concentration and fibre content on physic-mechanical properties of cement-bonded rattan fibre composites. Pro Ligno. Int J Wood Eng 11(4):192–198
41. Savastano (Jr) H, Warden PG, Coutts RSP, (2000) Brazilian waste fibres as reinforcement for cement-based composite cement. Concrete Comp 22:279–384
42. Tonoli GHD, Santos SF, Joaquim AP, Savastano (Jr) H (2010) Effect of accelerated carbonation on cementitious roofing tiles reinforced with lignocellulosic fibre. Constr Build Mater 24:193–201
43. Savastano (Jr) H, Warden PG, Coutts RSP, (2003) Potential of alternative fibre cement as building materials for developing areas. Cement Concrete Compos 31:232–243
44. Jayasankar R, Mahindran N, Ilangovan R (2010) ëStudies on concrete using Fly Ash, Rice Husk Ash and Eggshell Powderí. Int J Civ Struct Eng 1(3):362–373
45. ABNT (1993) Associacao Brasileira de Normas Tecnicas. NBR -7581. TelhaOndulada de Fibrocmento. Rio de Janerio: ABNT; Portuguese
46. Olorunnisola AO (2012) Looking beyond the challenges of affordable housing development, capitalizing on the engineering oppoutunities. The 18th Engr. Oluwemimo Arokodare lecture delivered on the 27th August, 2012, held at University of Ibadan Conference Centre, Ibadan
47. Okonkwo UN, Odiong IC, Akpabio EE (2012) The effect of eggshell ash on strength properties of cement-stabilized lateritic. Int J Sustain Constr Eng Technol 3(1):8–25

Non-destructive Test Approach for Evaluating High Strength Concrete Incoporating with Palm Oil Fuel Ash

Muhd Sidek Muhd Norhasri, Che Abdullah Fahim Aiman, Jumahat Aidah, A. H. Norhayati, H. Nuradila Izzaty, Newman Aidan, and Mohd Fauzi Mohd Afiq

Abstract Concrete is the main and popular construction material in Malaysia. Cost effectiveness and durability are the main factors that make them the best and affordable materials in Malaysia. In this paper, we present a non-destructive approach for high strength concrete with the inclusion of POFA. Apart from that, crack monitoring of high strength concrete with POFA is also monitored. High strength concrete is designed for Grade 60 and the utilization of POFA acts as an additive from 2.5, 5 and 7.5% from cement weight. All specimens will undergo water curing and testing on 7 and 28 days. All samples will be evaluated for NDT testing using rebound hammer and UPV and eventually after NDT samples will be tested for bending using an universal testing machine to see the flexural behaviour and crack pattern. From this report, it is seen that NDT testing shows a similar result as compared to the destructive test. In addition, the utilisation of POFA at every level of addition in concrete also enhanced the cracking behaviour of concrete. In conclusion, the utilisation of POFA is a proof to enhance the concrete performance in mechanical properties. Apart from that, the NDT approach can be an alternative to investigate the mechanical properties as compared to the destructive test.

Keywords Palm oil fuel ash · Non-destructive test · Rebound hammer · Ultrasonic pulse velocity · Cracking · Bending

---

M. S. Muhd Norhasri (✉) · J. Aidah · A. H. Norhayati
Institute for Infrastructure Engineering and Sustainable Management (IIESM), Universiti Teknologi MARA (UiTM) Shah Alam, Selangor, Malaysia
e-mail: norhasri@gmail.com; muhdnorhasri@uitm.edu.my

C. A. Fahim Aiman · H. Nuradila Izzaty · N. Aidan · M. F. Mohd Afiq
College of Engineering, School of Civil Engineering, Universiti Teknologi MARA (UiTM) Shah Alam, Selangor, Malaysia

M. Jawaid et al. (eds.), *Proceedings of 2nd World Conference on Byproducts of Palms and Their Applications*, Springer Proceedings in Materials 19,
https://doi.org/10.1007/978-981-19-6195-3_12

1 Introduction

Concrete is defined as a composite material, consisting mainly of Portland cement, water and aggregate. Concrete is the most commonly used man-made material on earth. Concrete is widely used from past until now and has a particular significance for structural engineers. It is an important construction material used in buildings, bridges, roads and dams.

Concrete is always used as the main building material because it has many benefits compared to other materials. First of all, concrete is economical. The production cost of cement concrete is very low. It is inexpensive and easily available around local areas compared to steel, polymers and other construction materials. The ingredients of concrete which are cement, water and aggregates also easily available in local markets at low cost. Next, concrete has the ability to be cast into any shape. Fresh concrete is in liquid state and can flow. Concrete can be poured into various formworks or shuttering configurations to form desired shapes and sizes at the construction site. Concrete can also be cast into complex shapes and configurations by adjusting the mix. Moreover, concrete is strong, durable and is of low maintenance. Concrete can lasts decades longer than other building materials getting stronger over time. This can reduce the total cost of the ownership as well as the environmental impact associated with more frequent rehabilitation and reconstruction. Furthermore, concrete is high-temperature resistance. Concrete doesn't burn, rust or rot. Concrete can withstand high temperatures better than wood and steel. In case of fire incidents, concrete can withstand for more than two hours to enable time for rescue missions in case of fire incident [1].

However, concrete also has a disadvantage that has become a concern to the engineers and researchers. The major problem of concrete is it has low tensile strength compared to compressive strength. Because of that, many concrete technologies were developed to counter this problem. There have been numerous experiments conducted by introducing a new material or recycled material to replace the aggregate or cement in concrete.

Nowadays, recycled materials tend to be used as concrete ingredients because there are increasing environmental legislations and people have started to develop awareness about the importance of environmental sustainability [2]. There is significant research on many materials to be used as cement substitute such as pulverized fuel ash (PFA), palm oil fuel ash (POFA) and other fibres. Since Malaysia is a major producer of palm oil, the wastage during palm oil production can be used to replace a small amount of cement.

Palm oil industry is one of the important industries in Malaysia. It is one of the important industries that generated Malaysia's economy. Malaysia is currently leading in the world as the the best palm oil producer. Malaysian Palm Oil Council (MPOC) stated that Malaysia carries out 39% of the world's palm oil production and 44% of the world's exports. However, the production of crude palm oil produces a large number of solid wastes. Solid wastes such as palm fibre, shells and empty fruit bunches are the output of crude palm oil production. Then, these wastes are burnt as fuel to generate electricity for extraction of palm oils. As a result, about 5% palm oil fuel ash (POFA) by weight of solid waste is produced. In Malaysia, utilization of palm oil fuel ash (POFA) is minimal and unmanageable, while its quantity increases annually. Most of POFA are disposed in landfills causing environmental and other problems. Many researchers have studied the use of agro waste ashes as materials in concrete, such as bagasse ash [3] and sawdust ash [4]. The results showed that these ashes contained large amounts of silica with little cementing properties. However, it can react with calcium hydroxide in the presence of moisture if it has a fine particle size to provide the cementing property. POFA is one of the agro waste ashes that contains large amount of silica and can be used as a cement replacement.

Non-destructive test (NDT) or non-destructive evaluation (NDE) was designed to examine an object without impairing its future usefulness, that inspects or measures without harm [5, 6]. This method was used to reduce the damage on the surface structure compared to destructive tests such as drilling cores. It can also be considered useful and valuable because it can control and monitor the manufacturing process, reduce the production cost and sustain the concrete's quality level. Several types of non-destructive testsinclude Schmidt Rebound Hammer, Ultrasonic Pulse Velocity (UPV), Half Potential Cell, XRF and Ground Penetrating Radar (GPR). All NDT methods have their own uses, such as rebound hammer, to determine the concrete's compressive strength, while the UPV is used to determine the quality of the concrete by using the time of longitudinal waves over a known distance [5–7].

The need for consistency characterization of deteriorated concrete structures provides NDT a considerable scientific and functional significance. The value of these measures can also be seen in the need for a potential change of use or structure expansion, acceptance of a structure to be purchased or insured, durability or integrity evaluation of repairs, and observing the performance of strengths related to curing, pre-stressing or loading.

2 Methodology

The preparation of raw materials is discussed briefly in this section. Ordinary Portland Cement for this study was procured by YTL Sdn. Bhd. Coarse aggregates were sieved using a 10 mm sieve. Meanwhile, the 4.25 mm sieve used to remove the silt. To maintain the freshness of the UHPC mix, GleniumAce 8538 was applied. POFA was heated up for 24 h and sieved under a 212 micro meter sieve. The raw materials are then poured inside the empty container and mixed by using a high speed mixer. Superplasticizer and water were poured slowly during the mixing process. The fresh concrete was then poured inside the beam mould of size 500 mm × 100 mm × 100 mm and vibrated to remove air bubbles. After 24 h it was demoulded and the air cured until the age of testing which was 7 and 28 days.

The performance of the concrete was tested by using the Non- Destructive Test (NDT) soon after the HPC was prepared. 1 out of 3 HPC specimen was used to determine the performance of the concrete. An indirect method was used when conducting the UPV test. Before placing the transducers on the surface of the beam, grease was applied at specific points. Reading of young modulus, time travel, effective length, compressive strength and speed displayed by the device were recorded. The sample is then tested with the Rebound hammer at a vertical position. The rebound number was then compared with Schmidt graph to determine the compressive strength. NDT test only involves only 1 beam which was cured for 7 and 28 days.

Crack monitoring is achieved by using 3 point the bending test. Before conducting the test, samples were applied with white paint and 50 mm × 50 mm grid lines were drawn. It helps in determining the cracks especially for a faint hair line crack. Crack monitoring only involving 2 beams which have been cured for 28 days.

3 Result and Discussion

Figure 1 portraits the compression strength for HPC with the inclusion of POFA by using the rebound hammer test. The test was conducted at age 7 and 28 days, respectively, and the HPC POFA was compared to the plain HPC. The inclusion of POFA can be seen effectively starting at day 7 and was marked by POFA HPC at 7.5% from cement weight as compared to the other mix. At day 7, the highest compressive strength was recorded by mix POFA 7.5% and followed by POFA 5%, OPC and POFA 2.5% with 38.27 MPa, 36.2 MPa, 36 MPa and 33.73 MPa in strength. Enhancement in strength performed by POFA 7.5% as compared to the mixes was contributed by the finer elements of POFA which was recorded at around 5-micron meter and thus performed a micro-filler effect in the microstructure of the concrete. This action created a denser surface between the cement binder where less void was created [1, 7]. The strength development was continued until age 28 where POFA 7.5% showed the optimum strength as compared to other mixes. The POFA 7.5% marked strength at 52.53 MPa followed by POFA 5% with 49 MPa, OPC with 47.7 MPa and POFA 2.5% with 35.8 MPa in compressive strength. The later action of POFA in enhancing the strength was embarked by the action of a supplementary cementitious material (SCM) with the elements of calcium carbonate. This element of calcium carbonate later on creates calcium oxide which produced additional hydration products by creating calcium silica hydrate (CSH). This finding was supported from the regression analysis (R^2) which differentiates the effect of POFA at day 7 and day 28 and shows that the inclusion of POFA as an early strength effect as compared to the later strength by R^2 marked at day 7 which is 0.87 as compared to R^2 at day 28 marked at 0.61. Thus supports the findings on the early strength effect by POFA.

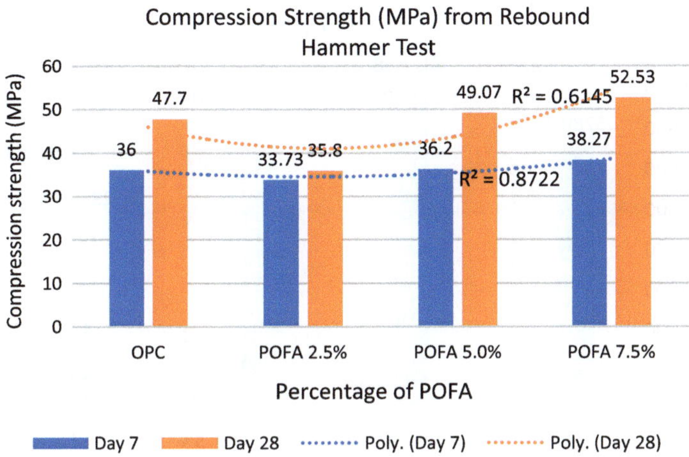

Fig. 1 Compressive strength for POFA HPC evaluated using Rebound Hammer Test

The quality of concrete for POFA HPC was evaluated using the UPV test at the age of 7 and 28 dayss and potrayed in Figs. 2 and 3. The result of UPV for POFAHPC shows a different perspective as compared to the rebound hammer. Day 7 shows that the best concrete was performed by POFA 2.5% and the lowest UPV value was recorded by mix POFA 7.5%. This action was contributed by the way the wave performed by UPV penetrate the concrete samples. In rebound hammer, the best mix was performed by POFA 7.5% but the reading for UPV was the lowest as compared to the other mixes. For rebound hammer, the impact of only touch at the surface but not for UPV which the wave can penetrate deeper. A different approach of testing shows that it was very important to combine several NDT approaches in concrete structures rather than relying on one NDT method. In addition, the higher the percentage of POFA recorded moderate reading of UPV and this was contributed by the interaction between the cement particles and POFA which might not fully uniform at a higher percentage as compared to a lower percentage of POFA and also the control specimen [8, 9]. This action was contributed by a smaller size of binding formation which was in nanoform and needs to be further investigated to see the formation of hydration products of POFA in nanoscale. Apart from that, the findings in Fig. 2 were supported by time travel or wave velocity values in Fig. 3. POFA HPC shows the lowest wave velocity as compared to the other samples. This shows that the inclusion of POFA performed a denser effect which that refines the microstructure of the concrete especially in the HPC elements [10, 11].

The performance of POFA HPC as structural components were derived in the crack versus load testing by using a rectangular beam in flexural strength test as shown in Fig. 4. Obviously, the inclusion of POFA increases the capacity of the HPC to carry maximum load. The optimum load was recorded by POFA 7.5% followed by POFA 5%, POFA 2.5% and OPC with 19.28 kN, 18.72 kN, 18.5 kN and 12.45 kN, respectively. From this the potential of POFA as a supplementary cementitious

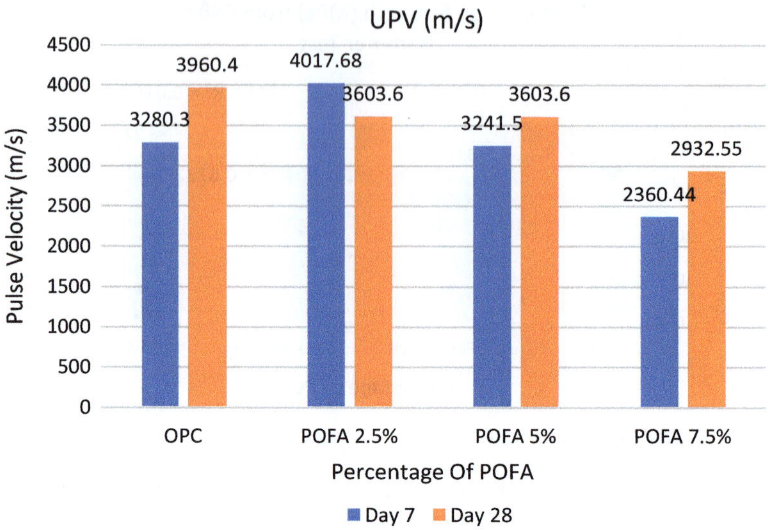

Fig. 2 Quality concrete for POFA HPC evaluated using Ultrasonic Pulse Velocity (UPV)

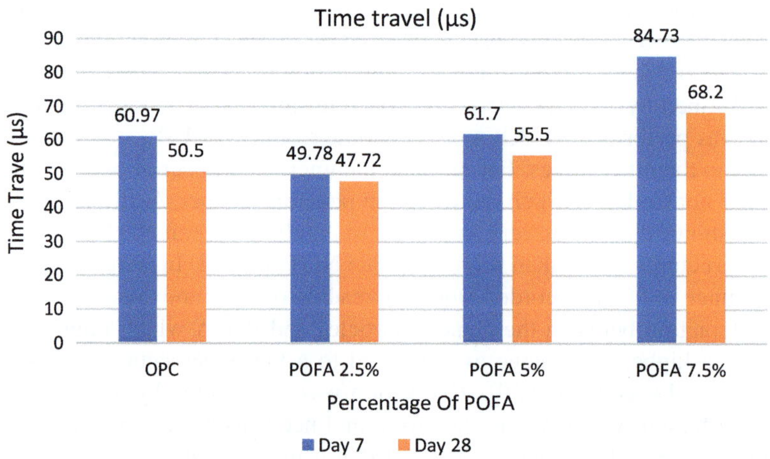

Fig. 3 Time travel for POFA HPC

material (SCM) is confirmed, which enhanced the performance in additional strength to the required HPC design. The action of SCM in POFA created an additional binding with the existing cementitious material [1, 12]. In this research the procurement of POFA was detected to be high in calcium which triggered the binding capabilities with cement and improved the bonding between cement and aggregates particles. It is also expected that the increase in the percentage of POFA will enhance the performance in flexural strength. Apart from the action of SCM, the size of POFA

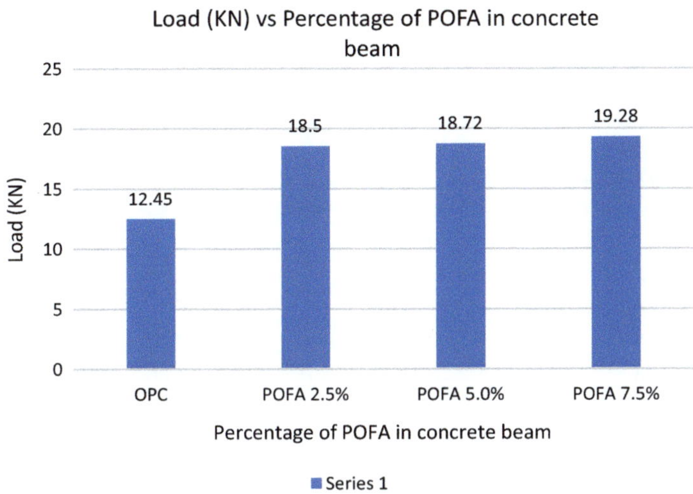

Fig. 4 Load versus POFA HPC

which was in round shape with a rough surface produced an additional bonding with cement along with the mobility of the POFA to fill in the voids between the cementitious particles [13, 14].

Figure 5 shows the crack displacement for POFA HPC as compared to OPC. It is confirmed that the inclusion of POFA as SCM in HPC reduce the crack pattern in the concrete beam. POFA 7.5% recorded the lowest reduction as compared to the POFA 5%, POFA 2.5% and OPC with 1.83 cm, 2.3 cm, 2.45 cm and 3.45 cm in length. The reduction of crack length performed by POFA HPC shows that the POFA has the potential of a fibre with elastic properties which can withstand the tension zone failure [15, 16]. In addition, the inclusion of POFA can also prolong the lifespan of the concrete by limiting the crack failure of the concrete. The enhancement in binding capabilities from the additional action of SCM by calcium from POFA proved to reduce the action of failure and the concrete crack length.

Figure 6 shows the relation between the load versus crack for inclusion of POFA in HPC. From this graph it is seen that the inclusion of POFA at every percentage level improved the capacity of the concrete beam to allow a high load. Furthermore, there was a proportional increase in the load for performance enhancement with reduction in crack length. The optimum relation to relate between the load versus crack was marked by mix POFA 7.5% followed by mix POFA 5%, POFA 2.5% and OPC with the relation of 19.28 kN and 1.83 cm, 18.72 kN and 2.24 cm, 18.5 kN and 2.45 cm and 12.45 kN and 3.45 cm in load and crack length, respectively. This relation was significantly confirmed by regression analysis for crack length that recorded R^2 value at 0.9 as compared to the regression value performed by the load with R^2 value at 0.69. The relation which shows that inclusion of POFA as a cement replacement material improved the crack distribution of the concrete beam as compared to the control sample. This was attributed by the action of SCM

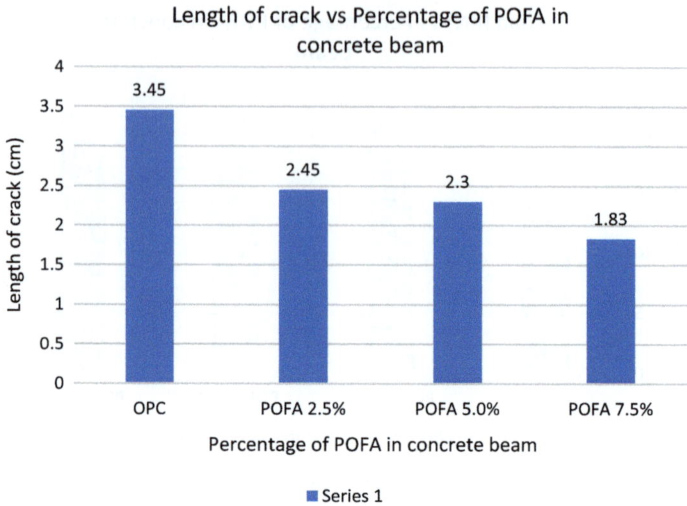

Fig. 5 Crack displacement for POFA HPC

in producing a good binding with cement particles. In addition, the reduction in crack length is proportional to the increase in load that shows the effect of ductility which produced concrete beam additional performance in withstanding the failure [17]. This action produced an alarm or warning on the behaviour of the beam due to failure. The reduction in deformation of the concrete beam also extended and the action of the POFA producing fibre oriented approach occurred in the hydration gel [18].

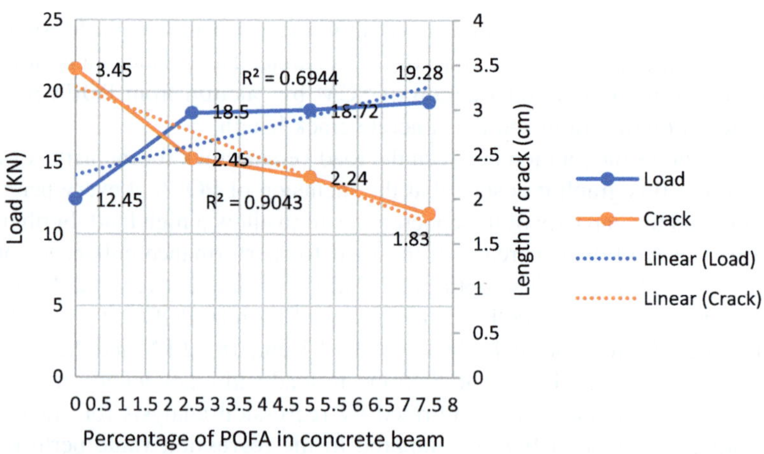

Fig. 6 Load versus crack for POFA HPC

4 Conclusion

From this research the following conclusions can be drawn :

a. Inclusion of Palm oil fuel ash (POFA) in High performance concrete (HPC) enhanced the workability of the concrete by producing a natural admixture.
b. The utilisation of a non-destructive test (NDT) such as the rebound hammer and ultrasonic pulse velocity produce an informative approach on the effectiveness of POFA as a cement replacement material in HPC.
c. The inclusion of POFA in HPC enhanced the performance of the concrete beam in producing a higher load.
d. Crack deformation of POFA in HPC was also enhanced by reducing the crack length of the concrete beam by the action of the fibre oriented approach.

Acknowledgement The author want to thank the Institute for Infrastructure Engineering and Sustainable Management (IIESM), Universiti Teknologi MARA (UiTM) and Ministry of Higher Education for providing the Fundamental research grant scheme (FRGS) Racer entitled 'Mechanism of mechano chemical analysis in transforming palm oil fuel ash (POFA) and waste paper sludge ash (WPSA) (600-IRMI/FRGS-RACER 5/3 (059/2019)).

References

1. Hamada HM et al (2021) Sustainable use of palm oil fuel ash as a supplementary cementitious material: a comprehensive review. J Build Eng 40:102286
2. Wasim M, Ngo TD, Law D (2021) A state-of-the-art review on the durability of geopolymer concrete for sustainable structures and infrastructure. Constr Build Mater 291:123381
3. Jannat N et al (2021) Utilisation of nut shell wastes in brick, mortar and concrete: a review. Constr Build Mater 293:123546
4. Pereira DC, Amaral-Labat G, Lenz e Silva GFB (2019) Effect of sawdust as porosity agent on final properties of geopolymers. Ceramica 65:104–109
5. Li D, Zhou J, Ou J (2021) Damage, nondestructive evaluation and rehabilitation of FRP composite-RC structure: a review. Constr Build Mater 271:121551
6. Muhd Norhasri MS, Shafee HM, Nurliza J, Mohd Afiq MF, Anizahyati A, Rohana H, Norhayati AH, Dzulkafley AS (2021) Evaluation of structural integrity for high rise buildings using non-destructive test method. Civil Eng Architect 9(5):12
7. Norhasri MSM et al (2021) Inclusion of palm oil fuel ash (POFA) as micro engineered material (MEM) in ultra high performance concrete (UHPC), in Lecture Notes in Civil Engineering, pp 57–66
8. Washer G et al (2005) Ultrasonic measurement of the elastic properties of Ultra High Performance Concrete (UHPC). In: Nondestructive evaluation and health monitoring of aerospace materials, Composites, and Civil Infrastructure IV. 2005. San Diego, CA
9. Owaid HM, Hamid RB, Taha MR (2017) Variation of Ultrasonic Pulse Velocity of multiple-blended binders concretes incorporating thermally Activated Alum Sludge Ash. KSCE J Civ Eng 21(4):1235–1246
10. Tambichik MA et al (2019) Combining pozzolanic material as ternary binder in Recycled Aggregate Concrete (RAC) to develop a new sustainable concrete. In: 2nd Global Congress on Construction, Material and Structural Engineering 2019, GCoMSE 2019. 2020. Institute of Physics Publishing

11. Bunnori NM et al (2020) relationships between compressive strength and transport properties of ultrahigh-strength green concrete utilizing ternary-blended binder. J Mater Civil Eng 32(3)
12. Elahi MMA et al (2021) Improving the sulfate attack resistance of concrete by using supplementary cementitious materials (SCMs): a review. Constr Build Mater 281:122628
13. Siddika A et al (2021) State-of-the-art-review on rice husk ash: a supplementary cementitious material in concrete. J King Saud Univ Eng Sci 33(5):294–307
14. Tamanna K et al (2020) Utilization of wood waste ash in construction technology: a review. Constr Build Mater, 237
15. Shaladi RJ et al (2020) Compressive strength and microstructural characteristics of binary blended cement mortar containing palm oil fuel ash. Lecture Notes in Civil Engineering. Springer, pp 1513–1522
16. Adebayo Mujedu K, Ab-Kadir MA, Ismail M (2020) A review on self-compacting concrete incorporating palm oil fuel ash as a cement replacement. Constr Build Mater 258:119541
17. Razali AW et al (2020) Optimization of palm oil fuel ash (POFA) content as cement replacement based on strength and porosity. In: Materials science forum, pp 3–7
18. Wan Hassan WNF et al (2019) engineering properties of high strength blended concrete enhanced with Nano POFA. In: 11th curtin university technology, science and engineering international conference, CUTSE 2018. 2019. Institute of Physics Publishing

Workability and Performance of High-Performance Concrete by Using Palm Oil Fuel Ash as a Cement Replacement Material

Muhd Norhasri Muhd Sidek, Mohamad Haris Hakim Mohd Nasir, Aidah Jumahat, Nuradila Izzaty Halim, Aidan Newman, Mohd Afiq Mohd Fauzi, and Nor Hayati Abdul Hamid

Abstract Malaysia has been one of the world's largest palm oil exporters which have been dealing with the issue of disposal of palm oil mill by-product known as palm oil fuel ash (POFA) over the years. However, over the past few years of research, this agro waste by-product has the potential to be utilized as construction material where it can be an alternative to conventional Ordinary Portland Cement (OPC). In the research, POFA is used as an alternative material to partially replace cement in concrete. The POFA used is treated to obtain the size of 212 microns before being used to partially replace cement by 0, 2.5, 5.0 and 7.5% of the cement weight. Workability properties of the concrete is covered by performing a slump test and flow table test. Whereas, the compressive and flexural test is conducted to obtain the mechanical properties of the concrete for curing age 1, 3, 7 and 28 days. It is revealed that the slump value increases as the POFA replacement percentage increases, however, the flow table shows a contradictory result. On the other hand, concrete with POFA shows enhanced mechanical properties, especially a 5% replacement percentage which is the optimum replacement level. Therefore, it is concluded that the inclusion of POFA in concrete thus improves the properties but only at an optimum level of replacement to cement.

Keywords POFA · Workability · Compressive · Flexural

M. N. M. Sidek (✉) · A. Jumahat · N. H. A. Hamid
Institute for Infrastructure Engineering and Sustainable Management (IIESM), Universiti Teknologi MARA (UiTM) Shah Alam, Selangor, Malaysia
e-mail: muhdnorhasri@uitm.edu.my

A. Jumahat
e-mail: aidahjumahat@uitm.edu.my

N. H. A. Hamid
e-mail: norha454@uitm.edu.my

M. H. H. M. Nasir · N. Izzaty Halim · A. Newman · M. A. M. Fauzi
School of Civil Engineering, Universiti Teknologi MARA (UiTM) Shah Alam, Selangor, Malaysia

M. Jawaid et al. (eds.), *Proceedings of 2nd World Conference on Byproducts of Palms and Their Applications*, Springer Proceedings in Materials 19,
https://doi.org/10.1007/978-981-19-6195-3_13

1 Introduction

Concrete is the most crucial and commonly used material in construction. Concrete is an environmentally friendly construction material which offer the stability and flexibility in designing all kinds of unique structures. Other than that, concrete has many impressive attributes such as fire resistance, high compressive strength and low maintenance. Nevertheless, concrete is prone to be brittle in behavior [1].

Other than that, concrete is also known for its major issue associated with its low tensile strength. Due to that, various new technologies and modern concrete specifications approaches were introduced [2]. Since then, there have been numerous experimental works conducted by introducing a new material or recycled material as a replacement to aggregate or cement in concrete.

Currently, concrete utilized with various types of recycled materials as part of its ingredient has gained in popularity and development due to the strict environmental legislation. Additionally, there are many research on many different materials to be used as cement substitutes and replacement such as palm oil fuel ash (POFA), pulverize fuel ash (PFA) and many other fiber and pozzolanic material [3]. Therefore, since Malaysia is the second largest palm oil industry producer, abundant wastage of palm oil can be used for this kind of research.

Palm oil industry is one of the most significant agro industries in Malaysia. Other than the production of crude palm oil, palm oil industry also produces a large amount of solid waste from the production process. Moreover, more than two million tonnes of solid waste of palm oil residues, such as palm fiber, shells and empty fruit bunches are produced once a year [4]. Utilization of POFA is minimal and uncontrollable as the quantity increases annually and most of the POFA are disposed of as waste in landfills which causes environmental problems. Despite that, many researchers has been studied on the use of waste ashes as constituents in concrete such as rice-husk ash, sawdust ash and bagasse ash [5, 6]. The results show that these waste ashes contained a high amount of silica in an amorphous form which could be used as a pozzolanic material. According to ASTM C 618 (2001), pozzolanic material is defined as a material that contains siliceous or siliceous and aluminous material by composition.

Therefore, a pozzolanic material may have little or no cementing properties. Though, when it has a fine particle size, it can react with calcium hydroxide at ordinary temperatures in the presence of moisture to provide the cementing property. POFA is one of the waste ashes which contains a large amount of silica and has a high potential to be used as a cement replacement [7].

According to the pioneer in POFA, Arif and Asrah [8] research acknowledged that POFA is a pozzolanic material and able to be partially replace cement up to 35% in mortar mix that could show similar strength as control mortar. Their studies have been continued by Mohamad and Mahmood [9] which highlighted that POFA concrete could gain maximum strength when 30% of the cement was replaced with POFA. It is reported that the maximum strength improvement occurred at the replacement level of 30% but a further increase in the ash content would result in the reduction of concrete strength gradually [10]. Besides that, increasing in the fineness of POFA would lead to greater concrete strength development than the coarser one due to an increase in density [11]. Of all the research that has been conducted, the performance of POFA incorporated as partial cement replacement in concrete still remains unexplored.

Most of the concrete produced nowadays is a multi component product which contain one or more admixtures in addition to the four basic ingredients which are cement, water, fine aggregate and coarse aggregate. For each and every component, one usually has several choices that could influence the cost, the concrete performance as well as the construction process. However, among the constituent ingredients, cement or cementitious materials as a whole play an important role in producing strong and durable concrete.

2 Methodology

Palm oil fuel ash (POFA) is a waste product obtained in the form of ash. In this study, the raw POFA has collected from a local factory processing palm oil in Nibong Tebal, Pulau Pinang. The collected POFA were dried in the oven at the temperature of 110 °C ± 5 for 24 hours before it can be used. Materials used in this experimental work consist of Ordinary Portland cement (OPC), coarse aggregate, fine aggregate, palm oil fuel ash (POFA), superplasticizer (SP) and water. Ordinary Portland cement conforming to MS 522: Part 1 for Portland cement specification from a single source was used as a binder. Coarse aggregate is obtained mostly from rock quarries. For this research, coarse aggregate size passing 10 mm and retained 5 mm according to ASTM C33 is used. Natural sand was used as fine aggregate with sieve sizing passing 2 mm and retained 600 μm in order to remove the bigger size of aggregate and impurities. Palm oil fuel ash (POFA) used was collected from a local palm oil factory located in Nibong Tebal, in the state of Pulau Pinang. The collected ashes were oven-dried at the temperature of 110 °C ± 5 for 24 hours before being sieved by passing through a 212 μm sieve. Superplasticizer classified as Type A according to ASTM 494-05 was also employed in all concrete mixes to improve the workability of concrete. Tap water was used for concrete preparation work and curing purposes.

To achieve the objectives stated, this experiment will use a modified design mix which was originally created by Norhasri and Hamidah [12] which utilizes the usage of nano metakaolin. For this experiment, metakaolin will be replaced with POFA with ascending order percentage of cement weight.

Table 1 shows the design mix formulation that will be used throughout in this research. After all the materials for the mix are prepared, a high speed mixer is later used to mix up all the materials together to allow homogeneity of the mix. The fresh concrete will be tested with a slump test and flow table test. All the mix will be cast into cube and beam samples to allow it to harden through the water cure. After the sample has hardened, it will be tested on compressive and flexural strength respectively.

Table 1 Quantities of the constituents per meter cubic

Mixture	Cement (kg/m^3)	Fine (kg/m^3)	Coarse (kg/m^3)	Superplasticizer (kg/m^3)	Water (kg/m^3)	POFA (kg/m^3)
OPC	800	433	800	16	160	–
P2.5	780	433	800	16	160	20
P5.0	760	433	800	16	160	40
P7.5	740	433	800	16	160	60

3 Result and Discussion

The workability for POFA HPC is shown by means of the slump test in Table 2 and Fig. 1 respectively. From the findings shows the inclusion of POFA in HPC increased the workability effect of HPC at every level of percentage. The highest slump readings were contributed by 7.5% replacement of POFA and followed by 5%, 2.5% and OPC with 310mm, 300mm, 290mm and 280mm in height. The inclusion of POFA gradually increased the workability by the action of shape and surface texture of the POFA itself. The shape of POFA was in rounded shape and the surface was rough and hard which potentially performed a ball bearing effect during mixing with the cement particles [13]. Combination of these two effects improved the uniformity of the concrete mix and the mobility of cement particles was improved. The enhancement of workability was needed in producing a uniform and well mixed concrete which later will improve the hardened properties of concrete. Furthermore, the enhancement of HPC with the aid of POFA proved the potential of POFA as a natural admixture. Since in this mix, the dosage of SP was fixed but at every percentage of POFA, the workability improved. It was believed that fineness, rounded shape and the hard surface texture were the main elements in producing the enhancement in workability [14]. Figure 1 shows the distribution of slump reading with the inclusion of POFA.

The hardened properties of POFA in HPC was evaluated in the compressive strength test and shown in Table 3 and Fig. 2. The enhancement in compressive strength can be seen at every level of percentage and the optimum effect in compressive strength was marked by 7.5% of POFA as cement replacement. The effect of POFA can be seen at an early age on day 1 when 7.5% of POFA recorded the highest strength at 67.35 MPa as compared to the control specimen OPC with 38.41 MPa. The strength gradually increased until day 28 when 7.5% POFA was the highest and followed by 5%, 2.5% and OPC with compressive strength of 89.02 MPa, 80.31

Table 2 Slump test results

Percentage of POFA (%)	Slump (mm)
0	280
2.5	290
5.0	300
7.5	310

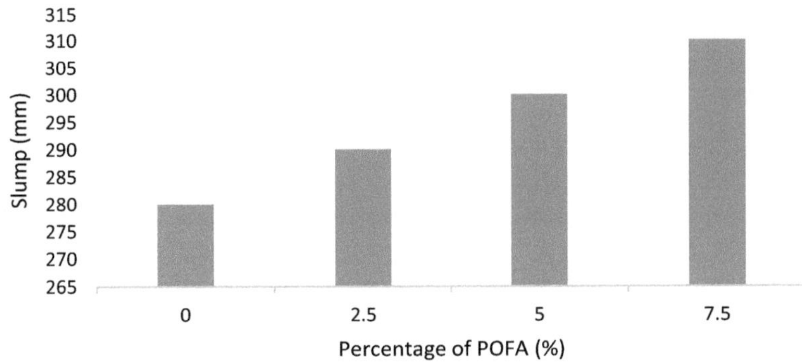

Fig. 1 Slump test of POFA HPC

MPa, 75.08 MPa and 68.6 MPa respectively. This effect was a portrait in Fig. 2 and shows a gradually increased compressive strength by means of POFA. The strength enhancement of POFA as cement replacement was attributed in 3 phases. First as filler agent, which refined the microstructure of HPC. Second as supplementary cementitious material (SCM) provided by additional calcium in the POFA which creates a good hydration product with cement. Thus, the additional calcium in POFA rapidly reacts with cement and accelerates the hydration process at early ages. Finally, as pozzolanic materials which were contributed by silica elements in the POFA whereby the strength enhancement was prolonged from this action [15]. These three actions were the elements that created the enhancement in strength. The uniqueness of POFA was the elements of calcium and silica were comparable with other reactive pozzolan and also SCM. Since POFA have both elements, the action of SCM and pozzolan can benefit in improving the HPC [16].

The flexural strength for POFA in HPC is shown in Table 4 and Fig. 3. In Table 4 shows a gradually increasing flexural strength for every level percentage of POFA. The early strength effect performed by POFA in HPC can be seen at day 1 respectively. The highest flexural strength was recorded by 2.5% POFA and followed by 5% POFA, 7.5% POFA and control specimen with 10.22MPa, 8.32MPa, 6.97MPa and 6.71MPa. The enhancement in flexural strength shows a similar pattern in day 3 with 2.5% POFA performing the optimum effect in strength. The early strength effect for POFA in HPC was contributed by the calcium elements in the POFA which performed supplementary cementitious materials (SCM) with cement. This action creates additional hydration products and at the same time denser the cement paste formation by performing as a filler.

The enhancement of flexural strength continued and can be seen in Fig. 3, especially at 7 and 28 days of testing. The optimum flexural strength for both 7 and 28 days was performed by 2.5% POFA and the lowest was the control specimen. The inclusion of POFA at 2.5% recorded the highest flexural strength at day 28 and followed by the control specimen, 5% POFA and 2.5% POFA with 18.96MPa, 17.86MPa, 16MPa and 14.58MPa respectively. The enhancement in flexural strength was attributed by the

Table 3 Compressive strength results for HPC with the inclusion of POFA

Type of concrete	Age of curing (day)	Average strength (N/mm^2)
Control sample	1	38.41
	3	45.46
	7	59.06
	28	68.6
2.5% of POFA	1	48.88
	3	50.63
	7	64.88
	28	75.08
5.0% of POFA	1	50.75
	3	52.99
	7	68.69
	28	80.31
7.5% of POFA	1	67.35
	3	59.67
	7	81.79
	28	89.02

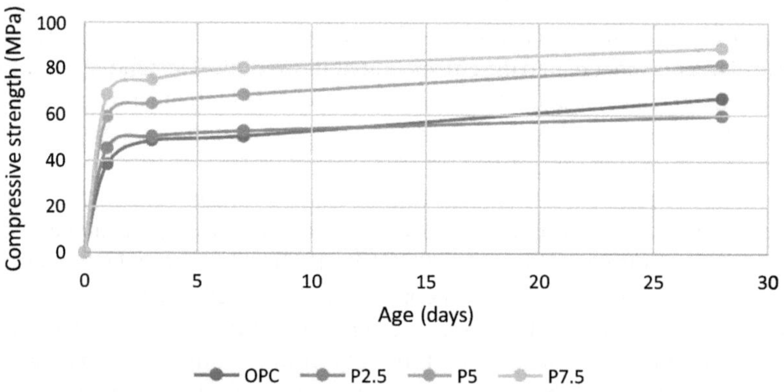

Fig. 2 Compressive strength test for POFA HPC

action of pozzolanic reaction performed by elements of silica in POFA which created the enhancement in the hydration process. the continuous action of pozzolanic reaction was prolonged until day 28 which was marked by 2.5% POFA as optimum mix in providing high flexural strength as compared to the other mix.

Table 4 Flexural Strength Results

Type of concrete	Age of curing (day)	Average strength (N/mm^2)
Control sample	1	6.71
	3	9.39
	7	11.87
	28	17.86
2.5% of POFA	1	10.22
	3	13.82
	7	16.20
	28	18.96
5.0% of POFA	1	8.32
	3	11.41
	7	14.16
	28	16.00
7.5% of POFA	1	6.97
	3	9.96
	7	12.32
	28	14.58

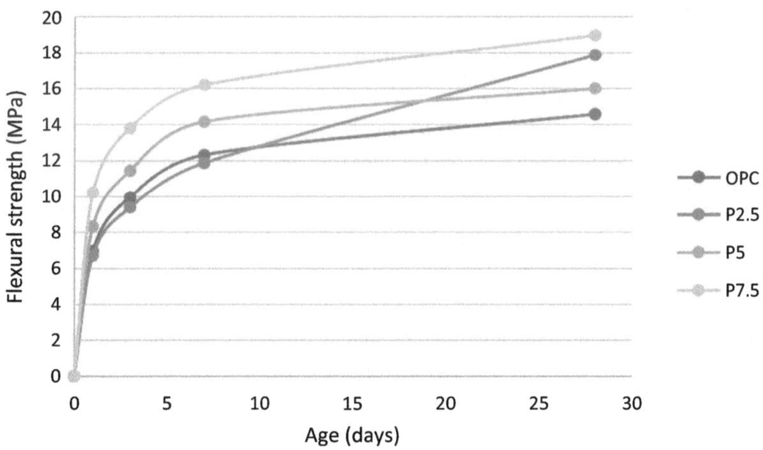

Fig. 3 Flexural strength of HPC with the inclusion of POFA

4 Conclusion

From this research conclusion can be drawn are:

a. The finer particle of POFA performs a filler effect and also enhance the workability of HPC by performing ball bearing effect when blends with cement

b. Inclusion of Palm oil fuel ash (POFA) in High performance concrete (HPC) enhanced the compressive and flexural strength of concrete by acting as supplementary cementitious material (SCM) and pozzolanic material.

Acknowledgements Author wants to thank the Institute for Infrastructure Engineering and Sustainable Management (IIESM), Universiti Teknologi MARA (UiTM) and Ministry of Higher Education for providing Fundamental research grant scheme (FRGS) Racer entitled 'Mechanism of mechanochemical analysis in transforming palm oil fuel ash (POFA) and waste paper sludge ash (WPSA) (600-IRMI/FRGS-RACER 5/3 (059/2019)).

References

1. Hamada HM et al (2018) The present state of the use of palm oil fuel ash (POFA) in concrete. Constr Build Mater 175:26–40
2. Paiva H et al (2017) Microstructure and hardened state properties on pozzolan-containing concrete. Constr Build Mater 140:374–384
3. Hamada HM et al (2021) Sustainable use of palm oil fuel ash as a supplementary cementitious material: a comprehensive review. J Build Eng 40:102286
4. Wahab N et al (2021) Performance of palm oil fuel ash (Pofa) in peat soil stabilization. Malaysian Constr Res J 13(SUPPL.2):197–211
5. Hamada HM et al (2020) Effect of high-volume ultrafine palm oil fuel ash on the engineering and transport properties of concrete. Case Stud Constr Mater 12:e00318
6. Salami BA et al (2019) Modelling the early strength of alkali-activated cement composites containing palm oil fuel ash. Proc Inst Civil Eng Constr Mater 172(3):133–143
7. Shakir A et al (2019) Characterization of palm oil fuel ash as cementitious supplement: a review. ACI Mater J 116(6):31–44
8. Arif A et al (2019) The properties of blended cement containing palm oil fuel ash. In: 11th curtin university technology, science and engineering international conference, CUTSE 2018. 2019. Institute of Physics Publishing
9. Mohamad ME et al (2016) A review of the mechanical properties of concrete containing biofillers. In: International engineering research and innovation symposium, IRIS 2016. 2016. Institute of Physics Publishing
10. Safiuddin M, Salam MA, Jumaat MZ (2011) Utilization of palm oil fuel ash in concrete: a review. J Civ Eng Manag 17(2):234–247
11. Ali N et al (2016) Potential mixture of POFA and SCBA as cement replacement in concrete—a review. In: 2016 international symposium on civil and environmental engineering, ISCEE 2016. 2017. EDP Sciences
12. Norhasri MSM, Hamidah MS, Fadzil AM (2019) Inclusion of nano metaclayed as additive in ultra-high-performance concrete (UHPC). Constr Build Mater 201:590–598
13. Abdul Awal ASM et al (2017) Mechanical properties and thermal behaviour of two-stage concrete containing palm oil fuel ash. Int J GEOMATE 12(32):166–175
14. Runyut DA et al (2018) Microstructure and mechanical characterization of alkali-activated palm oil fuel ash. J Mater Civil Eng 30(7)
15. Adebayo Mujedu K, Ab-Kadir MA, Ismail M (2020) A review on self-compacting concrete incorporating palm oil fuel ash as a cement replacement. Constr Build Mater 258:119541
16. Li Z, Lu D, Gao X (2021) Optimization of mixture proportions by statistical experimental design using response surface method—a review. J Build Eng 36:102101

Other Applications of Palm Byproducts

Utilization of Oil-Palm Leaves for Making Innovative Products: A Comprehensive Review

Arif Nuryawan and Iwan Risnasari

Abstract To add value to oil palm (*Elaeis guineensis* Jacq.) leaves (OPL) agricultural residue, this study aimed to explore the use of this lignocellulose material for making innovative products. Traditionally, OPL has been only utilized as cattle feed. In this study, five innovative products made of OPL have been presented, namely (i) particleboard; (ii) sheathing products; (iii) comply board; (iv) insulation board; and (v) eco-printing products. Particleboard has been made from OPL using 10% urea–formaldehyde (UF) resin based on dry leaves with two different amounts of hardeners, namely 1% and 3% ammonium chloride. Sheathing products were similar to particleboard with some enhancements of raw material, for instance, increasing of hardener amount if still using UF resin; replacing UF with exterior type adhesive like isocyanate; and layering thin veneer on both surfaces of the board. Comply board has had performance similar to plywood because of its face/back layers using thicker wood veneer; thus it can be used as a structural application. Insulation board made of OPL mixed with recycled paper as the raw material was able to absorb sound. Testing of its capability involved acoustical property. Eco-printing on fabric and paper using OPL only resulted in patterns without any colors. Eco-printing is a combination of dyeing and printing through natural colorants existing in plants such as leaves including OPL.

Keywords Particleboard · Sheathing products · Comply board · Insulation board · Eco-printing products

1 Introduction

There are three kinds of biomass that have been generated during oil palm (*Elaeis guineensis* Jacq.) growing in plantations, consisting of oil palm trunk (OPT), empty

A. Nuryawan (✉) · I. Risnasari
Faculty of Forestry, Department of Forest Products Technology, Universitas Sumatera Utara, Indonesia and Member of AEPI (Association of Eco-Printer of Indonesia), Medan 20155, North Sumatra, Indonesia
e-mail: arif5@usu.ac.id

M. Jawaid et al. (eds.), *Proceedings of 2nd World Conference on Byproducts of Palms and Their Applications*, Springer Proceedings in Materials 19,
https://doi.org/10.1007/978-981-19-6195-3_14

fruit bunches (EFB), and oil palm fronds (OPF) with its leaflet [1]. The use of OPT for various products has been extensively explored and some good results have been reported, ranging from engineered to advanced products such as conventional panel products (plywood and laminated products) [2, 3], composite products (impregnated products and composite plastics) [4, 5], and carbohydrate derivatives (starch/sap/sugar) [6–8]. Similarly, EFB also has been utilized to a large extent using a biorefinery concept. A review on its utilization concluded that EFB can be converted into many variants of bio-products and chemicals, such as polymers, pharmaceuticals, chemical building blocks, biofuels, and bioenergy [9]. OPF in particular has been only used as a roughage source or as a component part in complete feed for livestock [1] even though there are possibilities for making alternative lumber [10] or carbon sources [11] thus it can support wood manufacturers and biochemical industry, respectively.

When OPF (Fig. 1) was utilized as raw material for biocomposite products, i.e., lumber substitution [10], or feedstock of biofuel [11], i.e., succinic acid via a fermentation process, the leaves were separated from the frond, underutilized only for cattle feed or discarded around plantation site.

Leaves of OPF are pinnate and reach between 3 and 5 m long. A young palm can produce 30 leaves per year, meanwhile mature palm with the age of over 10 years can generate about 20 leaves annually [12], therefore oil palm leaves (OPL) is one of the potential biomass residues from agricultural crops. In Southeast Asia, particularly in Indonesia, Malaysia, and Thailand, OPL are availability abundant since the area of this plantation increased extensively in a decade as shown in Fig. 2.

In Indonesia, generally, OPL has been only utilized as ruminant feed [14], compost [15], and bio-mulch [16] because of its non-toxicity, higher nutrient content, and economically practical, respectively. Therefore, in recent years a lot of research focused on the fundamental properties of the OPL, including the production rate of OPL in wet and dry areas [17], measurement of growth leaf area for photosynthesis study [18], or prediction on nutrient content within the OPL [16, 19].

Further, several studies have shown that OPL contained minor components which exhibit beneficial properties such as anti-oxidants compounds [20] and anti-microbial activity thus suitable for cosmetics [21], anti-bacterial activity thus its active ingredient potential for soap formulation [22, 23], cellulose microfiber (CMF)/microfibrillated cellulose extract for probiotic encapsulant [24], flavonoids and cathechin

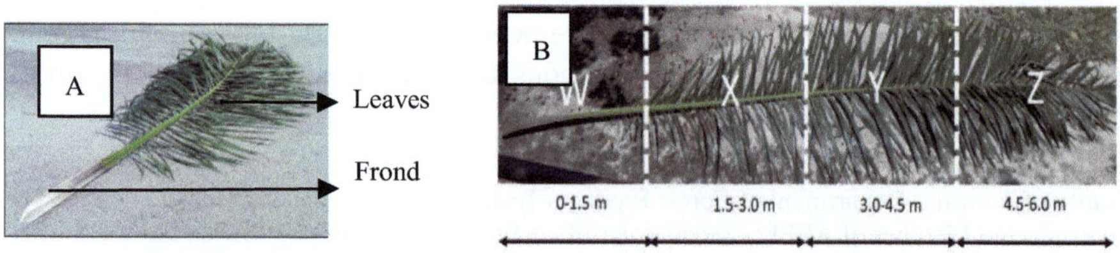

Fig. 1 **a** OPF and its leaflet (from Ref. [10]) and **b** segmentation of OPF and leaves, W part about 1.5 m long is no leaves (from Ref. [11])

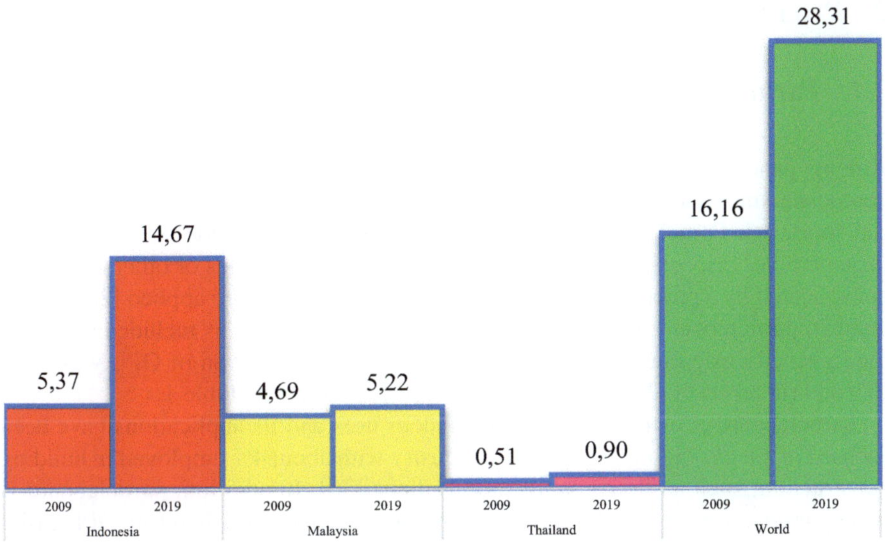

Fig. 2 World area of oil palm harvested (green) within the predominant three countries in Southeast Asia during 2009–2019; consisting of Indonesia (red), Malaysia (yellow), and Thailand (pink) (in million hectars) (own elaboration according to Ref. [13])

thus effective for diabetes complication treatment [25], anti-tumor, anti-cancer, anti-proliferative and anti-inflammatory properties thus potential to be medicinal purposes [26] and food functional ingredient [27].

Despite the aforementioned extensive research work on the development of OPL, all these studies employed only a small amount of OPL. In other words, only small parts, i.e., extracts, active compounds and potential ingredients, of OPL have been utilized. Application research involving voluminous OPL biomass seemed underdeveloped and had poor attention. Therefore, this review is intended to more explore the potential of OPL for product manufacturing made of a large quantity of OPL biomass. In this review, five innovative products made of enormous OPL have been presented, namely particleboard, sheathing products, comply board, insulation board, and eco-printing products. The first four are common worldwide because the methodology for making these products are available elsewhere in published papers. However, this review will emphasize in their functions such as packaging and furniture substrate (for particleboard); wall, ceiling panels, office dividers, bulletin board, cabinet, counter/desktop (for sheathing products); light construction and structural functions (for comply board); and sound absorption (for insulation board). The last, eco-printing products will require enormous OPL if the production rate is massive in a such region. In this context, OPL could be utilized as a source of natural dyes as well as patterns form. Each product was described not only in its characteristics but also technically how it was prototyped and made in the laboratory.

2 Products Made of Oil Palm Leaves

2.1 Particleboard

Among panel products, particleboard has been the most popular conventional composite since the production rate of plywood decreased because of raw material shortages. Particleboards are conceptualized as panels made from lignocellulosic particles, especially wood particles, or and synthetic resins or other adhesives, consolidated by a process in which temperature and pressure are applied [28]. In this context, particleboard made of OPL is feasible, because OPL is included as one of the lignocellulosic materials. Further, the chemical composition of OPL was close to wood (in terms of alpha cellulose and lignin) as shown in Table 1.

Particleboard is usually intended for indoor uses and its applications have been focused in the packaging and furniture industry with about 4% employed in building construction, such as floors, walls, and ceilings [30]. In addition, particleboard is widely used as a core for thin overlays in the furniture industry, therefore OPL easily and provides a promising medium for making this substrate [31]. In other words, particleboard made of OPL has not been ready for direct uses, it has to be layered, veneered, coated, or laminated. In addition, particleboard made of OPL showed lower physical and mechanical properties [29, 31]. Overlays onto both surfaces will enhance these properties. Discussion of this topic will be provided in the next sub-heading related to comply product.

Some factors influenced the quality of particleboard made of OPL, such as raw material, adhesive and hardener used, technical procedure, and preparation specimens for testings. In our case [32], leaves were depicted without mixing with other parts like petiole and rachi as shown in Fig. 3.

Because of interior purposes, urea–formaldehyde (UF) resin usually was used as the adhesive of the particleboard. Fortunately, UF resin is cured in acid condition or lower pH, therefore ammonium chloride (NH4Cl) was added to the glue prior to applying within the furnish. Either 1% or 3% of NH4Cl was added for adhesive curing referred to previous work [33]. Our student has experimented to apply 1 and 3% of NH4Cl to OPL's particleboard with 8, 10, and 12% UF resin adhesive content based on oven-dry OPL [34]. Results of this work showed that OPL's particleboard with higher adhesive content and hardener, viz. 12% UF and 3% NH4Cl, has had better

Table 1 Chemical composition of OPL comparing to wood (from Ref. [29])

Name	Extractives	Chemical composition (%)		
		Holocellulose	α-cellulose	Lignin
OPL	20.60	47.70	44.53	27.35
Hardwood	0.1–7.7	71–89	31–64	14–34
Softwood	0.2–8.5	60–80	30–60	21–37

Fig. 3 Parts of OPL, for particleboard raw material in our study [32], only leaflets were considered as raw material. Petiole and rachi have been removed because both belonged to OPF (from Ref. [16])

properties comparing with the others. Performance of the resulted particleboard has shown in Fig. 4.

Both technical procedure and sample's preparation for particleboard evaluation were referred to Indonesian Standard of SNI 03–2105-2006 [35] and Japanese Industrial Standards (JIS) A 5908: 2003 [36], respectively.

Fig. 4 Performance of OPL's particleboard bonded by UF resin (from Ref. [34])

2.2 Sheathing Products

Product developments of particleboards were making them for sheathing applications such as floor, wall, ceiling panels, bulletin boards, cabinets, furniture, countertops and desktops [37]. One of the prerequisites of these applications is higher compaction among particles within the particleboard. For reaching this, the internal bonding of the particles within the board should be strong. However, our students' works revealed that bonding OPL's particles within the OPL's particleboards were less compact thus worsening their physical and mechanical properties [32, 34]. Therefore, some recommendations for their enhancements should be conducted, for instance addition of more hardeners, application of surface layer, or replacement interior type adhesive into exterior one [38].

In addition more hardener was intended for making less viscous of the UF resin's adhesive thus the resin could distribute thoroughly because OPL was voluminous. Unfortunately, as our work employed three kinds amount of hardener, namely 5%, 7% and 9%; the study revealed that higher hardener resulted in lower mechanical properties [39] even though the performance looked similar as shown in Fig. 5.

Addition of the aforementioned hardeners dropped pH values to 7, 5 and 3, respectively [38]. Indeed, the higher amount of hardener resulted in lower pH which created a more acidic condition for making UF cured faster. However, this situation was unbeneficial for the mechanical properties of the particleboard. Acidity could destroy the structure of lignocellulose [40] thus weakening the bonding among OPL's particles. Further, UF resin itself will degrade when it is exposed to acid [41]. It was believed that the application of layers onto both face and back surfaces improved particleboard properties, including their dimensional stability [42]. Therefore, wood veneers were placed onto OPL's particleboard surfaces as shown in Fig. 6.

Another factor affecting the quality of OPL's particleboard when used as sheathing products is adhesive type. It is obvious that interior type adhesive is inferior compared to exterior type adhesive, for instance, amino resin like thermosetting UF adhesive has had less power compared to phenol and isocyanate adhesive after application in wood bonding because of hydrolytic degradation reason [44]. Therefore, alteration using

Fig. 5 Performance of OPL's particleboard bonded by UF resin with different amounts of hardeners, namely 5%, 7% and 9% (from Ref. [38])

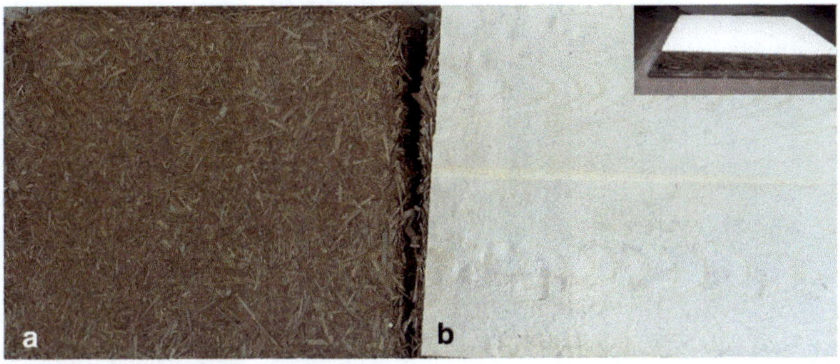

Fig. 6 OPL's particleboard becomes a core in between wood veneers on face and back layers; **a** without layers, **b** with thin veneer layers, and insert from side view (from Refs. [39, 43])

isocyanate for keeping strong OPL's particleboard was conducted [39]. Isocyanate has been used as particleboards' adhesive in recent years, and the type of MDI or 4–4'-diphenyl methane diisocyanate has been used nowadays [45, 46]. When the isocyanate was used as adhesive in particleboards system, its level content can be lower because of its high reactivity. Attempts have been made to use isocyanate for binding leaves particleboard [39, 46].

Figure 7 showed a comparison of typical leaves particleboard bonded with lower content of isocyanate.

Fig. 7 Performance of leaves particleboard, OPL's particleboard versus mahagony leaves particleboard (from Refs. [39, 46])

Fig. 8 Bending testing of veneered OPL's particleboard using one point loading of UTM (from Ref. [43])

2.3 Comply Board

Comply was defined as three- or five-layers panels with a special type of plywood with particleboard as the core and used for construction [47]. In this context, comply board was related to loading and intended for structural application, particularly light construction. Some decorative features on both surfaces (face and back layers) could be originated from the coating, plastics, or wood veneers [36].

Thickness of wood veneers influenced OPL's particleboard properties, particularly their mechanical properties [43]. Therefore, there is a specific classification of veneered particleboards in JIS standards according to their bending strength, namely type of 30–15, which means the minimum value of modulus of rupture (MoR) testing depends on the direction of wood veneer grain, for example, 30 N/mm^2 for perpendicular and 15 N/mm^2 for parallel, respectively [36].

Testing of MoR is carried out together with the modulus of elasticity (MoE) using a Universal Testing Machine (UTM). Similar with MoR, the value of MoE also depended on the direction of wood veneer grain, which resulted in a value minimum of 4000 N/mm^2 for lengthwise and 2800 N/mm^2 for widthwise. Typical investigation on MoR and MoE of veneered OPL's particleboard was shown in Fig. 8.

2.4 Insulation Board

At first, the insulation board was referred to as a group of fiber panel products manufactured to specific gravities ranging from about 0.16–0.50 for providing thermal protection to the wall construction and serving to reduce noise transmission through the exterior walls [47]. Because there was a development of insulation board made of nonwood particleboard [48, 49], there was also a possibility for making insulation board from OPL. In this case, OPL's particleboard was emphasized only for acoustic

Fig. 9 Typical performance of sound insulation board made of OPL and recycle paper (from Ref. [50])

purposes since tropical countries did not have a winter season and thus rarely need a heat absorber for making indoors warm. Therefore, in the manufacturing process, recycle papers were added into the furnish in order to make a good sound absorber. Typical performance of the board was presented in Fig. 9.

Indeed, acoustical properties related to control sound isolation/sound insulators and sound absorption/noise absorbers, both have distinct characteristics. The first is related to sound waves transmitted between rooms in a building, depending on material properties, and expressed as the transmission loss factor or the ability of the material to isolate a sound. The latter is correlated to the sound quality in a room and expressed as the sound absorption coefficient [48]. Both parameters were measured using the impedance tube method [50] and several factors such as density, porosity, and thickness influenced resulted values.

2.5 Eco-Printing Products

Eco-printing is an art of combining dyeing and printing through natural colorants existing in plants, fruits, vegetables, waste materials, or by-products [51]. Attempts have been made for using OPL for producing astonishing and interesting shades, textures, traces and marks either on fabric surfaces or paper as shown in Fig. 10.

In our study, eco-printing using OPL resulted in a pattern without any colors. Here, color was derived from natural dye incorporated in the eco-printing process, i.e., bark of mangrove tree yielded in gradual colors of brown, red-blackish until

Fig. 10 Typical performance of OPL eco-printing on fabric and paper surfaces (from Refs. [52, 53])

brown-blackish [53]. Indeed, the visual of eco-printing products depended on the technique or process used, which usually consisted of at least three steps, namely mordanting, contact printing and dyeing, and fixation processing. Mordanting is a method for maintaining the basic color of the fibers within the fabric or paper thus opening their pores by immersing them in water containing metal at a certain time [51, 55]. Contact printing and dyeing were arranging sources of natural colorants such as leaves and other parts of plants according to the desired pattern onto the fabric or paper [54]. In this step, either natural color and pattern were extracted by boiling, steaming, beating/pounding, or combination among them. Prior to extraction, the fabric or paper was sandwiched between two media (fabric or plastic), rolled, wrapped, and fastened tightly for hindering penetration of water as well as undesired visual effects of the pattern. The fixation is a finishing process for binding motifs and colors that have been printed on fabric or paper [54]. A study reported that steaming fixation resulted in better color strength evaluated by reflectance technique compared to conventional fixation like direct boiling in hot water due to condensation of water vapour may accelerate the penetration of dye molecules into fabric or paper [55].

Eco-printing using OPL should be continued in terms of exploring its potency whether using old and young leaves, utilizing of wet and dry leaves or making a pattern with the basic form of OPL. Previous study on eco-printing work revealed that this activity was able to engage the community, resulting in many variations of products, and become a sustainable business [56, 57].

3 Conclusion

Added value of OPL can be reached by utilizing them into innovative products, for instance, conventional particleboard, sheathing, comply board, insulation board, and eco-printing products. OPL particleboards have lower physical and mechanical properties, therefore they should be layered using veneer, coating, or lamina. Further, UF resin as the binder limits OPL particleboard as only interior properties such as furniture or indoor applications. Differently, sheathing products required higher compaction among OPL particles, therefore more hardener was needed if the UF was applied. Another type of adhesive such as isocyanate can also be used, therefore the application of sheathing was more broad compared to OPL particleboard, not only for horizontal but also vertical installation. Comply board was special for structural purposes. It was related to loading work because the construction was similar to plywood. Only the core was made of OPL and the rest was veneers. Insulation boards were related to thermal protection or sound absorption; therefore, the composition of the panel was important. Addition of recycled paper can either absorb thermal or noise. Eco-printing products performed astonishing and interesting shades, textures, traces and marks either on fabric surfaces or paper originating from OPL with aid of mordant.

Acknowledgements We extend our sincere thanks to Universitas Sumatera Utara (USU) for funding this activity through *Penelitian Unggulan Universitas* (PUU) year of 2017 (for preliminary research) with Research Contract of TALENTA USU fiscal year of 2017 number 5338/UN5.1.R/PPM/2017 date of May 22, 2017 and *Program Pengembangan Usaha Produk Intelektual Kampus* (PPUPIK) multiyears of 2019-2021 (for application and commercialization) with contract number of 328/UN5.2.3.2.1/PPM/2019 (date of May 20, 2019), number of 288/UN5.2.3.2.1/PPM/2020 (date of June 9, 2020 for Batch I) and number of 174/UN5.2.3.2.1/PPM/2021 (date of May 25, 2021). Further, thanks also to Rumah Nauli Apik for developing an eco-printing product, ByPalma for recommending this paper into this journal, our students (Dame Lasmaria Siahaan, Jan Heri Damanik, Aidul Adhansyah, Melinda Rosanny Manalu, Harley Christian, and Elia Arimby Sinurat) for lab workings.

References

1. Zahari MW, Hassan OA, Wong HK, Liang JB (2003) Utilization of oil palm frond—based diets for beef and dairy production in Malaysia. Asian-Aust J Anim Sci 16(4):625–634
2. Feng LY, Tahir PM, Hoong YB (2011) Density distribution of oil palm stem veneer and its influence on plywood mechanical properties. J Appl Sci 11(5):824–831
3. Nordin K, Jamaludin MA, Ahmad M, Samsi HW, Salleh AH, Jalaludin Z (2004) Minimizing the environmental burden of oil palm trunk residues through the development of laminated veneer lumber products. Manag Environ Qual Int J 15(5):484–490
4. Rosli F, Ghazali CMR, Abdullah MMAB, Hussin K (2016) A review: characteristics of oil palm trunk (OPT) and quality improvement of oil palm trunk plywood by resin impregnation. BioResources 11(2):5565–5580

5. Lubis MJ, Risnasari I, Nuryawan A, Febrianto F (2009) The quality of composite board made of waste oil palm stem (*Elaeis guineensis* Jacq) and recycle polyethylene (PE). J Tek Ind Pert 19(1):16–20
6. Mansor H, Ahmad AR (1990) Carbohydrates in the oil palm stem and their potential use. J Tropical Forest Sci 2(3):220–226
7. Tomimura Y (1992) Chemical characteristics of oil palm trunk. Bull For Prod Res Inst 362:133–142
8. Dirkes R, Neubauer PR, Rabenhorst J (2021) Pressed sap from oil palm (*Elaeis guineensis*) trunks: a revolutionary growth medium for the biotechnological industry? Biofuels, Bioprod. Bioref. 15:931–944
9. Rame: Oil palm empty fruit bunches (OPEFB): existing utilization and current trends bio refinery in Indonesia. E3S Web of Conferences 31, 03014 (2018)
10. Rasat MSM (2014) Biocomposite lumber from oil palm frond: an alternative in wood-based industry from utilization of oil palm fronds agriculture residue. Universiti Malaysia Kelantan, 1–117
11. Tan JP, Jahim JM, Harun S, Wu TY, Mumtaz T(2016) Utilization of oil palm fronds as a sustainable carbon source in biorefineries. Int J Hydrogen Energy 41:4896–4906
12. Sumathi S, Chai SP, Mohamed AR (2008) Utilization of oil palm as a source of renewable energy in Malaysia. J Renew Sustain Energy Rev 12:2404–2421
13. [FAO] Food and Agriculture Organization. Forestry Production and Trade. Available online : http://www.fao.org/faostat/en/#data/FO. Accessed 25 Dec 2020
14. Rizali A, Fahcrianto, Ansari MH, Wahdi A (2018) Utilization of fresh waste and palm oil leaves through fermentation *Trichoderma* sp. as cut cow feed. EnviroScienteae 14:1–7
15. Daryono, Alkas TR (2017) Utilization of waste district and palm oil leaves. J Hutan Tropika 5:88–195
16. Mandang T, Sinambela R, Pandianuraga NR (2018) Physical and mechanical characteristics of oil palm leaf and fruits bunch stalks for bio-mulching. IOP conference series : earth and environmental science, vol 196, p 012015
17. Breure CJ (1994) Development of leaves in oil palm (*Elaeis guineensis*) and determination of leaf opening rate. Expl Agric 30:467–472
18. Gerritsma W, Soebagyo FX (1999) An analysis of the growth of leaf area of oil palms in Indonesia. Expl Agric 35:293–308
19. Santoso H, Tani H, Wang X, Segah H: Predicting oil palm leaf nutrient contents in Kalimantan, Indonesia by measuring reflectance with a spectroradiometer. Int J Remote Sens.https://doi.org/10.1080/01431161.2018.1516323
20. Han NM, May CY (2010) Determination of antioxidants in oil palm leaves (*Elaeis guineensis*). Am J Appl Sci 7(9):1243–1247
21. Yusof NZ, Gani SSA, Siddiqui Y, Mokhtar NFM, Hasan ZAA (2016) Potential uses of oil palm (*Elaeis guineensis*) leaf extract in topical application. J Oil Palm Res 28(4):520–530
22. Ahmad N, Hasan ZAA, Muhamad H, Bilal SH, Yusof NZ, Idris Z (2018) Determination of total phenol, flavonoid, antioxidant activity of oil palms leaves extracts and their application in transparent soap. J Oil Palm Res.https://doi.org/10.21894/jopr.2018.0010
23. Febriani A, Syafriana V, Afriyando H, Djuhariah YS (2020) The utilization of oil palm leaves (*Elaeis guineensis* Jacq.) waste as an antibacterial solid bar soap. IOP conference series : earth and environmental science, vol 572, p 012038
24. Pato U, Ayu DF, Riftyan E, Restuhadi F, Pawenang WT, Firdaus R, Rahma A, Surono IS, Jaswir I (2021) Physicochemical property of oil palm leaves and utilization of cellulose microfiber as probiotic encapsulant. Biodiversitas 22(7):2937–2944
25. Rajavel V, Sattar MZA, Abdulla MA, Kassim NM, Abdullah NA (2012) Chronic administration of oil palm (*Elaeis guineensis*) leaves extract attenuates hyperglycaemic-induced oxidative stress and improves renal histopathology and function in experimental diabetes. Evid Based Compl Alternat Med Article ID 195367:1–12
26. Owoyele BV, Owolabi GO (2014) Traditional oil palm (*Elaeis guineensis* Jacq.) and its medicinal uses : a review. TANG Humanitas Med 4(3):e16

27. Mohamed S (2014) Oil palm leaf : a new functional food ingredient for health and disease prevention. J Food Process Technol 5:300
28. Maloney TM (1992) Modern particleboard & dry-process fiberboard manufacturing. Miller Freeman, San Francisco
29. Hashim R, Nadhari WNAW, Sulaiman O, Kawamura F, Hiziroglu S, Sato M, Sugimoto T, Seng TG, Tanaka R (2011) Characterization of raw materials and manufactured binderless particleboard from oil palm biomass. Mater Des 32:246–254
30. Bertolinia MS, Lahra FAR, Nascimentoa MF, Agnelli JAM (2013) Accelerated artificial aging of particleboards from residues of CCB treated *Pinus* sp. and castor oil resin. Mater Res 16(2) :293–303
31. Hashim R, Nadhari WNAW, Sulaiman O, Sato M, Hiziroglu S, Kawamura F, Sugimoto T, Seng TG, Tanaka R (2012) Properties of binderless particleboard panels manufactured from oil palm biomass. BioResources 7(1):1352–1365
32. Nuryawan A, Rahmawaty, Damanik JH (2017) Characteristics of particleboard made from mahagony and oil palm leaves. Proceeding—5th Kuala lumpur international agriculture, Forestry and Plantation, pp.187, Hotel Bangi-Putrajaya, Bandar Baru Bangi Selangor, Malaysia, July 31–August 1, (2017)
33. Nuryawan A, Park B-D (2014) Comparison of thermal curing behavior of liquid and solid urea-formaldehyde resin with different formaldehyde/urea mole ratios. J Thermal Analyt Calorimetry 118:397–404
34. Siahaan DL (2017) Quality of particleboard made of oil palm (*Elaeis guineensis* Jacq.) leaves with variation levels of urea-formaldehyde (UF) adhesive. [Undergraduate thesis, in Bahasa Indonesia], Faculty of Forestry, Universitas Sumatera Utara
35. [BSN] *Badan Standardisasi Nasional,* Standar Nasional Indonesia (SNI) 03–2105–2006 for particleboards [in Bahasa Indonesia], Jakarta: BSN (2006)
36. [JSA] Japanese Standard Association, Japanese Industrial Standard (JIS) A 5908 : 2003 for particleboards, Japan : JSA (2003)
37. Wang D, Sun XS (2002) Low density particleboard from wheat straw and corn pith. Ind Crops Prod 15:47–50
38. Nuryawan A, Rahmawaty (2018) Properties of leaves particleboard for sheathing application. IOP conference series : earth and environmental science 126:012032
39. Adhansyah A (2017) Improvement of particleboard quality made of oil palm (*Elaeis guineensis* Jacq.) leaves [Undergraduate thesis, in Bahasa Indonesia] Faculty of Forestry, Universitas Sumatera Utara
40. Ruhendi S, Febrianto F, Sahriawati N (2000) Liquefied wood for exterior plywood adhesive [in Bahasa Indonesia] J Ilmu Pertanian Indonesia 9(1) :1–11
41. Nuryawan A, Park B-D (2014) Microstructure of cured UF resins modified by rubber latex emulsion after hydrolytic degradation. J Korean Soc Wood Sci Technol 42(5):605–614
42. Massijaya MY, Nuryawan A, Assyh N (2005) Fundamental properties of com-ply made of small diameter fast growing species and mersawa veneer. Proceedings of Scientific Session 90 XXII IUFRO World Congress), pp 86, Brisbane, August 12 (2005)
43. Manalu MR (2017) Enhancement of design and quality of particleboard made of oil palm leaves into comply products [Undergraduate thesis, in Bahasa Indonesia] Faculty of Forestry, Universitas Sumatera Utara
44. Nuryawan A, Park B-D (2017) Quantification of hydrolytic degradation of cured urea-formaldehyde resin adhesives using confocal laser scanning microscopy. Int J Adhes Adhes 74:1–5
45. Nuryawan A, Alamsyah EM (2019) Thermal properties of isocyanate as particleboard's adhesive. IOP conference series : materials science and engineering, vol 593, p 012004
46. Nuryawan A, Alamsyah EM (2019) Thermal stability of isocyanate as particleboard's adhesive investigated by TGA (Thermogravimetric Analysis). IOP conference series : earth and environmental science, vol 374, p 012004
47. Saota HJ (2017) Improvement of particleboard quality made of mahagony leaves [Undergraduate thesis, in Bahasa Indonesia] Faculty of Forestry, Universitas Sumatera Utara

48. Bowyer JL, Shmulsky R, Haygreen JG (2003) Forest products and wood science an introduction, 4th edn. Iowa State Press A Blackwell Publ, USA
49. Karlinasari L, Hermawan D, Maddu A, Martianto B, Lucky IK, Nugroho N, Hadi YS (2012) Acoustical properties of particleboards made from betung bamboo (*Dendrocalamus asper*) as building construction material. BioResources 7(4):5700–5709
50. Iswanto AH, Hakim AR, Azhar I, Wirjosentono B, Prabuningrum DS (2020) The physical, mechanical, and sound absorption properties of sandwich particleboard (SPb). J Korean Wood Sci Technol 48(1):32–40
51. Christian H (2020) Evaluation of acoustical properties of particleboard made of oil palm (*Elaeis guineensis* Jacq.) leaves. Paper presented at Faculty of Forestry, Universitas Sumatera Utara
52. Ismal ÖE (2016) Patterns from nature : contact printing. J Textile Assoc
53. Nuryawan A, Risnasari I (2021) Utilization of oil-palm leaves for making innovative products. Poster presented at ByPalma 21 conference, Kuala Lumpur, Malaysia
54. Sinurat EAB (2021) Utilization of bakau minyak (*Rhizophora apiculaca*) bark as color fixation on fabrics and paper with ecoprinting coloring. Universitas Sumatera Utara, Paper presented at Faculty of Forestry
55. Nurcahyanti D, Septiana U (2018) Handmade eco print as a strategy to preserve the originality of Ria Miranda's design in the digital age. MUDRA J Art Culture 33(3):395–400
56. Rekabya M, Salemb AA, Nassara SH (2009) Eco-friendly printing of natural fabrics using natural dyes from alkanet and rhubarb. J Textile Inst 100(6):486–495
57. Nuryawan A, Risnasari I, Irwansyah A, Pulungan W, Sinaga AS, Ginting FYE, Munthe M (2020) Eco-print on recycle paper/ fabrics as main products of business development program of intellectual property campus of Universitas Sumatera Utara. J Saintech Transfer (JST) 3(1):1–11

A New Approach for Studying the Dyeability of Date Palm Residues Fabric with Sustainable Natural Dyes

Noureddine Baaka and Ramzi Khiari

Abstract The date palm (*Phoenix Dactylifera*), found mainly in the United States (*California*), North Africa and the Middle East, plays a significant contribution to the environmental and economic conditions in these regions. In addition to their high nutritional food value, date palms provide a wide range of abundantly renewable agricultural by-products. The date palm tree supplies eight types of residues: rachis, spines, leaflets, bunch, fibrillium, pedicels, spathe and petiole, that can be harvested during the seasonal pruning stage. In Tunisia, date palm tree acquires great importance historically, socially and economically. The pruning, especially the leaves, are used in various traditional industries and construction by farmers and artisans. Currently, craftsmen used synthetic dyes to dye their products. This paper is devoted to develop a dyeing process of handicrafts products with natural dyes. The colour yielding plant materials obtained from natural dyes extracted from date palm were compared to the dyes extracted from various sources such as: henna (*Lawsonia inermis*) leaves, madder (*Rubia tinctoria*) roots and turmeric (*Curcuma Longla L.*) rhizome. It can be assumed that the dyes extracted from date palm using different mordants can offer a rich palette of colours in various shades and tones. The several mordant's effects (ferrous sulphate and alum) on the colour and fastness properties of the dyed samples were studied comparatively. The colour of the dyed samples was studied according to the CIELab values (L*, a* and b*) and K/S values. The results seem to be very interesting and encouraging to explore at industrial textile scale.

N. Baaka · R. Khiari (✉)
Laboratory of Environmental Chemistry and Clean Process (LCE2P- LR21ES04), Faculty of Sciences of Monastir, University of Monastir, 5019 Monastir, Tunisia
e-mail: khiari_ramzi2000@yahoo.fr

N. Baaka
Higher Institute of Fashion of Monastir, 5019 Monastir, Tunisia

R. Khiari
Department of Textile, Higher Institute of Technological Studies of Ksar Hellal, Ksar Hellal, Tunisia

Université Grenoble Alpes, CNRS, Grenoble INP, LGP2, F-38000 Grenoble, France

M. Jawaid et al. (eds.), *Proceedings of 2nd World Conference on Byproducts of Palms and Their Applications*, Springer Proceedings in Materials 19,
https://doi.org/10.1007/978-981-19-6195-3_15

Keywords Natural dyeing · Handicrafts · Date palm leaves · Fastness properties

1 Introduction

The date palm plays a very significant role in southern Tunisia, both from a socio-economic and ecological view point. It forms the basis of the economy of the Djerid and Nefzaoua regions and occupies a privileged position in the national economy. The area occupied by date palm reached about 40 976 ha, with a phoenicicoleheritage which has attained nearly 5.5 million palms of all varieties, including more than 66% (3.66 million) of Deglet Nour's feet. These areas are distributed among 54 415 plots of which 67.6% (14 716 ha) of traditional type and the rest of new plantations.

Although Tunisia's agricultural heritage is spread over four governorates, it is mainly concentrated in the southern governorates. Kébili is in first place with more than 55% of the national heritage, as is the case for the cultivated area.

The other Tunisian palm trees are located in Tozeur (21%), Gabes (16%) and Gafsa (5%). Tunisia is the world's leading exporter in terms of value. Tunisian dates capture 23% of the value of world trade.

In addition to its production of dates for human consumption, the date palm offers a wide range of by-products exploited by the Saharan population [1, 8] namely.

- Date juice, by extraction, is used as a sweetener;
- Tree trunk, used in traditional cabinet making, firewood and building frames;
- Dry palms, used as fences, windbreaks, in the manufacture of mats, hand fans, bread dishes, baskets, mats, various types of containers for domestic use and also as packaging material for fruit and vegetables (Fig. 1) [5],
- Date bunches, as traditional brooms, and as fuel;
- The life for the confection of the soles of sandals;

The date palm continues to make the news as researchers discover new ways to reuse its waste. Date palms by-products are used in many fields (Fig. 2). In fact, many studies are devoted to the use of date palm fibers as reinforcing elements in

Fig. 1 Photographs of handicraft products made from date palms

composite materials. Belgacem et al. [3] show in particular that the use of cellulosic fibers extracted from the date palms allows to obtain a material with good mechanical properties. During this work, date palm fibres were obtained by different process namely: chemical, enzymatic and mechanical treatments.

Furthermore, Riahi et al.[13] were studied the application of date palm fiber filters as a porous medium for tertiary treatment of domestic wastewater. According to this study, date palm fiber filtration could be a potential technology for tertiary wastewater treatment, as it offers a "green engineering solution".

In addition, the rachis was investigated, as reported by Khiari et al. [9, 10], for papermaking. It was concluded that from rachis date palm, an excellent paper with physical and chemical properties were obtained. On the other hand, date palm oil from Phoenix dactylifera seeds has been used for its antifungal and antioxidant activities [6, 7].

Everything in the palm tree is exploited from composed plants namely: the palm sticks, the leaves, the fruit and the drink made from it. So, palm trees are an important resource for people as well as for an economic society. In Tunisia, the use of palm leaves is widespread and sustains the lives of the respected families who live on the palm plantation. The people who manage the housework also work with these palm leaves. According to the customer's request, the product is woven and finished. This is more of a small-scale work section that aims to enhance living conditions and maintain the practice of palm leaf handicrafts. During this work, we are devoted to

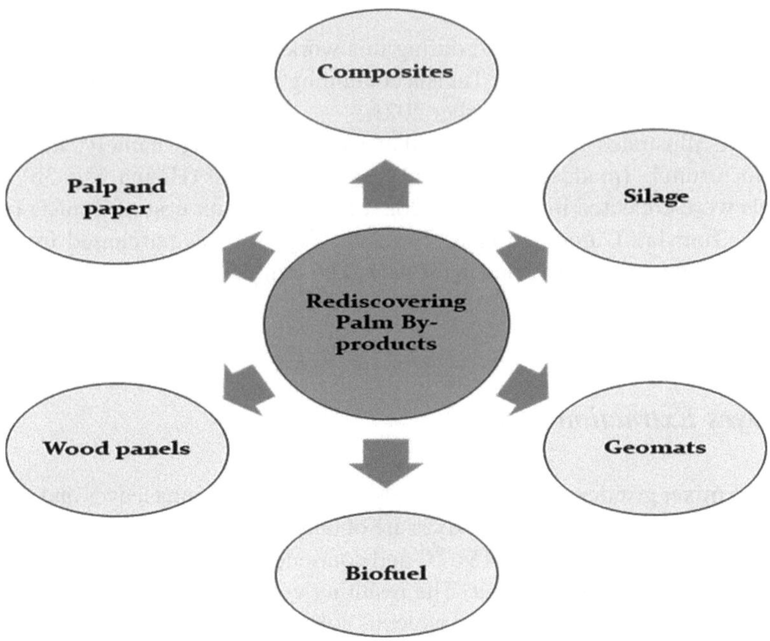

Fig. 2 Rediscovering date palm by-products

Fig. 3 (**a**) *Rubia tinctorum* L., (**b**) *Lawsonia inermis* and (**c**) *Curcuma longa*

develop a dyeing process of handicrafts products with natural dyes. Three natural dyes were used such as henna leaves, madder roots and turmeric rhizome.

2 Experimental

2.1 Materials

Date palm materials (leaves) used during this work were collected from the Gabes region (a city in the south east of Tunisia containing an oasis with many hundreds of thousands of palm trees) in December 2021.

Figure 3 illustrates the three dyes that were investigated namely: the species Rubia tinctorum L. (madder, Fig. 3a) and Lawsonia inermis (Henna, Fig. 3b). These materials were collected in December 2021, from the south-east of Tunisia (region of Gabes, Tunisia). Curcuma longa (Turmeric, Fig. 3c) was purchased from local market. All plant matter was dried in the dark. The samples were grinded into powder using a laboratory grinder.

2.2 Dyes Extraction Method

By using a mixer grinder, the three materials (madder root, henna leaves and turmeric rhizome) are ground and powdered dyes are obtained. 5 g of each sample along with 100 ml distilled water is heated at 90 °C and continuously stirred for 1 hour by using a hot plate with a magnetic stirrer. The resulting coloured solutions are filtered by plain cotton fabric and the extracted aqueous solutions are used for dyeing.

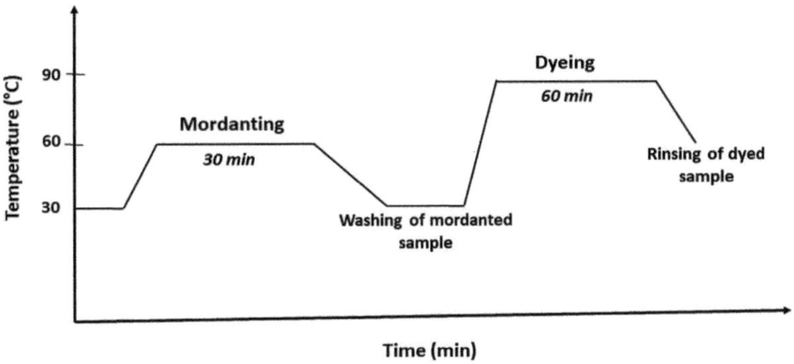

Fig. 4 Graphic representation of the mordanting and dyeing operation

2.3 Dyeing and Mordanting Method

The pre-mordanting method was used to mordant date palm leaves with 5% (w.w.) ferrous sulphate ($FeSO_4.7H_2O$) and 10% (w.w.) alum ($Al_2K_2(SO_4)_4.24H_2O$). The samples were submerged in mordant solutions at 30 °C with a liquor ratio of 40:1. The temperature was progressively risen to 60 °C at a rate of 2 °C per minute, and maintained for 30 min. The mordanted samples were carefully washed with tap water.

The control dyeing (without mordant) was carried out at a liquor ratio of 40:1 under dyeing conditions of 90 °C for 60 minutes at neutral pH. Finally, the samples were rinsed several times with tap water and dried in the shade. A graphical illustration of the mordanting and dyeing process is illustrated in Fig. 4.

2.4 Colour Parameters

The colour yield (*K*/*S*) is selected as mainly factor in estimating the dyeing quality [2]. It is calculated with a Datacolour 800 spectrophotometer. The *K*/*S* was then estimated using the following Kubelka–Munk equation [11, 12].

$$^{K}/_{S} = \frac{(1-R)^2}{2R} - \frac{(1-R_0)^2}{2R_0}$$

where R and R_0 are the decimal fraction of the reflectance of the stained and unstained sample, respectively, K is the absorption coefficient, and S is the scattering coefficient.

3 Results and Discussion

3.1 Colour Shades of Date Palm Leaves Dyed with Natural Dyes

Turmeric is a dye that is known for having incoherent colour changes with mordants [15]. This dye can be employed for dyeing into various shades of yellow by varying the selected mordants. Thus, the results presented in Table 1 which it show that alum yielded bright yellows and ferrous produced shades of yellow brownish. The colour values (K/S) of date palm leaves dyed with madder, henna and turmeric with and without pre-mordanting are presented in Fig. 5.

From this figure, it is concluded that the K/S values of the dyed leaves are higher in pre-mordanting with ferrous sulphate than with alum. From all the mordants, the fabric dyed with $FeSO_4$ has a maximum K/S value, which might be related to a higher affinity of iron and a higher colour depth that also changes the colour tone of the dye. The K/S values for pre-mordanting with various mordants were in the following order: $FeSO_4$>no mordant>aluminum. The collected results show that the colour strength of the dyed sheets is highly dependent on the ability of the metal mordant to

Table 1 Colour shades of date palm leaves dyed with natural extracts of madder, turmeric and henna under different conditions

Sample	Shades		
	Madder	Henna	Turmeric
Unmordanted			
5% $FeSO_4$			
10% Alum			

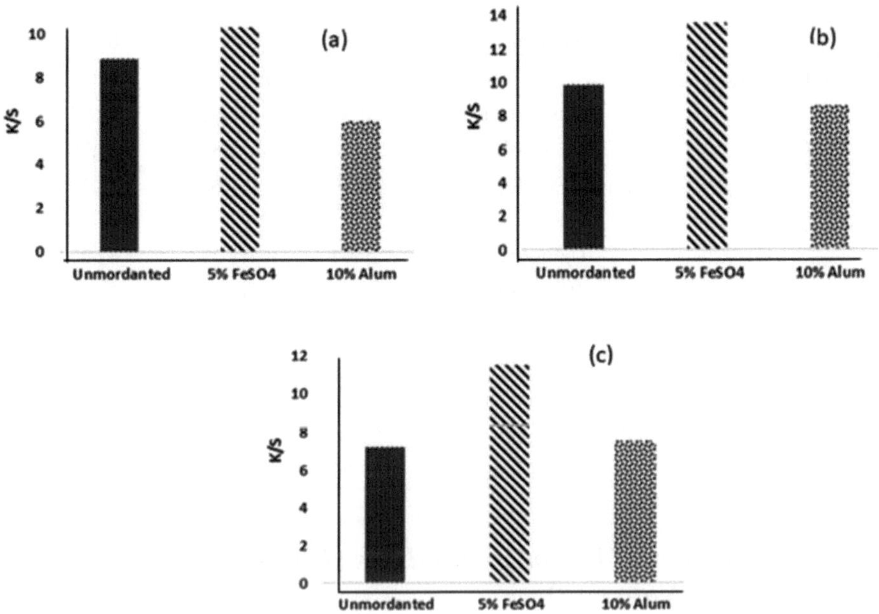

Fig. 5 Colour yield (K/S) values of unmordanted and mordanted dyed leaves with; (**a**) Madder; (**b**) Henna; (**c**) Turmeric

form the coordination complex. Ferrous sulphate can form high coordination bonds [4], while aluminum has weak interactions between the fabric, mordant and dye, and thus the absorption of the dye is comparatively lower [14].

3.2 *Colourimetric Properties of Unmordanted and Mordanted Dyed Leaves*

The (L, a* and b*) values of date palm leaves dyed with madder, henna and turmeric with and without pre-mordanting are presented in Table 2. According to this table, it may be noted that the mordants have higher and lower L-values, imparting lighter and deeper hues, respectively. Similarly, positive a* and negative b* represent yellow and red respectively.

Table 2 Colourimetric properties of unmordanted and mordanted dyed leaves

Sample	L*	a*	b*
	Madder		
Unmordanted	53.30	16.91	42.92
5% $FeSO_4$	27.62	5.89	15.70
10% Alum	36.29	19.62	38.20
	Henna		
Unmordanted	29.94	18.32	42.22
5% $FeSO_4$	25.94	19.21	37.53
10% Alum	32.89	20.25	43.52
	Turmeric		
Unmordanted	68.08	3.94	62.19
5% $FeSO_4$	57.35	9.17	53.16
10% Alum	67.52	4.59	71.07

3.3 Development of Prototype Products (Handicraft Hat) Dyed with Natural Dyes

Palm leaves are woven into a variety of things: boxes, display boxes, simple square boxes, hats, baskets and much more. After all the previous experiments in colouring the date palm leaves, the researcher will turn to the main core of the paper which is using these dyed palm leaves in making modern crafts according to the international fashion trends. Thus, two prototypes of handicrafts from date palm leaves were produced. The results are presented in Table 3.

4 Conclusions

The date palm offers a wide range of renewable and abundantly available agricultural by-products. These by-products are used in many traditional industries and construction by cultivators and craftsmen. During this paper, handicrafts products are dyed with three natural dyes namely i.e.; henna leaves, madder roots and turmeric rhizome. It is observed that a large colour palette of shades of varied hue and tone can be obtained by using different mordants. It has been concluded that sustainable shade developers (mordants) have improved fastness ratings.

Table 3 Development of two prototype products

Products	Descriptions	Pictures
Handicraft hat	Date palm leaves have been dyed with turmeric according to the process described in Sect. 2.3 The dyed leaves have been used to weave a handicraft hat	
Handicraft hat	Date palm leaves have been mordanted with ferrous sulphate. Then, the mordanted leaves have been dyed with madder according to the process described in Sect. 2.3 The dyed leaves have been used to weave a handicraft hat	

References

1. Anwar MA (2006) *Phoenix dactylifera* L.: a bibliometric study of the literature on date palm. Malays J Libr Inf Sci 11:41–60
2. Baaka N, Mahfoudhi A (2017) Orange peels waste as a low-cost cotton natural dye. Mor J Chem 5:259–265
3. Belgacem C, Serra-Parareda F, Tarrés Q, Mutjé P, Delgado-Aguilar M, Boufi S (2021) Valorization of date palm waste for plastic reinforcement: macro and micromechanics of flexural strength. Polymers 13:1751
4. Bhattacharya SD, Shah AK (2000) Metal ion efect on dyeing of wool fabric with catechu. Color Technol 116:10–12
5. Chao CCT, Krueger RR (2007) The date palm (Phoenix dactylifera L.): overview of biology, uses and cultivation. Hort science 42:1077–1082
6. Chibane E, Essarioui A, Ouknin M, Boumezzourh A, Bouyanzer A, Majidi L (2020) Antifungal activity of Asteriscus graveolens (Forssk.) Less essential oil against Fusarium oxysporum f. sp. albedinis, the causal agent of "Bayoud" disease on date palm. Mor J Chem 8:456–465
7. El Ouadi Y, Beladjila A, Bouyanzer A, Kabouche Z, Bendaif H, Youssfi F, Berrabah M, Touzani R, Chetouani A, Hammouti B (2017) The Palm oil from seed of Phoenix dactylifera (Oil of both Deglet Nour and Kentichi) as a natural antioxidants and Environment-Friendly inhibitors on the Corrosion of mild Steel in HCl 1M. Mor J Chem 5:139–152
8. Hassan SK, Bakhsh ZA, Gill A, Maqbool A, Ahmad W (2006) Economics of growing date palm in Punjab, Pakistan. Pakistan. Int J Agric Biol 8:788–792

9. Khiari R, Mhenni MF, Belgacem MN, Mauret E (2010) Chemical composition and pulping of date palm rachis and Posidonia oceanica – a comparison with other wood and non-wood fibre sources. Bioresour Technol 101:775–780
10. Khiari R, Mauret E, Belgacem MN, Mhenni F (2011) Tunisian date palm rachis used as an alternative source of fibres for papermaking applications. Bioresources 6:265–281
11. Kubelka P (1948) New contributions to the optics of intensely light-scattering materials. Part I. J Opt Soc Am 44:448–457
12. Kubelka P (1954) New contributions to the optics of intensely light-scattering materials. Part II. J Opt Soc Am 44:330–334
13. Riahi K, Ben Mammou A, Ben Thayer B (2009) Date-palm fibers media filters as a potential technology for tertiary domestic wastewater treatment. J Hazard Mater 161:608–613
14. Uddin MG (2014) Effects of different mordants on silk fabric dyed with onion outer skin extracts. J. Text 1–8
15. Vankar P, Shanker R, Wijayapala S, De Alwisb A, De Silva N (2008) Sonicator dyeing of cotton, silk and wool with Curcuma domestica Valet extract. Int Dyers 193:38–42

Activated Carbon from Date Palm Rachis for Continuous Column Adsorption of *o*-Cresol

Nisrine Khadhri, Manel Elakremi, Ramzi Khiari, and Younes Moussaoui

Abstract High surface area microporous activated carbon has been prepared from date palm rachis by chemical activation with hydroxide sodium. The process has been conducted at different impregnation ratios (NaOH/precursor = 0.5–4) and carbonization temperatures (500, 600 and 700 °C). The physical structure and chemical properties of obtained activated carbon were derived from Scanning Electron Microscope (SEM), N_2 adsorption/desorption isotherms, Fourier-transform infrared spectroscopy (FTIR), thermo gravimetric analysis (TGA), Boehm titration and pH zero point charge measurement (pH_{PZC}). The activated carbon obtained under optimal conditions (600 °C, RAM = 3, 2 h) has a mesoporous structure with a specific surface area of 1108 $m^2\ g^{-1}$ and its surface contains mainly basic groups with a $pH_{ZCN} = 8$. Activated carbon was used as an adsorbent for the removal of *o*-cresol from aqueous solutions in continuous mode. The four most popular breakthrough models, namely, Adams–Bohart, Thomas, Yoon–Nelson and Yan were used for the correlation of breakthrough curve data along with the evaluation of model parameters. The Bohart-Adams model describes admirably the initial part of the breakthrough curve ((C_t/C_0) < 0.5), and the hole curve was well fitted by the Yoon-Nelson and Thomas models.

N. Khadhri
Laboratory for the Application of Materials to the Environment, Water and Energy (LR21ES15), Faculty of Sciences of Gafsa, University of Gafsa, Gafsa, Tunisia

M. Elakremi · Y. Moussaoui (✉)
Organic Chemistry Laboratory (LR17ES08), Faculty of Sciences of Sfax, University of Sfax, Sfax, Tunisia
e-mail: y.moussaoui2@gmx.fr

Faculty of Sciences of Gafsa, University of Gafsa, Gafsa, Tunisia

R. Khiari
Laboratory of Environmental Chemistry and Cleaner Process (LCE2P-LR21ES04), University of Monastir, 5019 Monastir, Tunisia

Higher Institute of Technological Studies (ISET) of Ksar-Hellal, 5070 Ksar-Hellal, Tunisia

University of Grenoble Alpes, CNRS, Grenoble INP, LGP2, F-38000 Grenoble, France

M. Jawaid et al. (eds.), *Proceedings of 2nd World Conference on Byproducts of Palms and Their Applications*, Springer Proceedings in Materials 19,
https://doi.org/10.1007/978-981-19-6195-3_16

Keywords Activated carbon · Date palm · Chemical activation · Fixed-bed column · Breakthrough curve

1 Introduction

Activated carbon is the most widely used adsorbent for the removal of pollutants from wastewater. Activated carbons are known as very effective adsorbents due to their highly developed porosity and amorphous carbonaceous material of variable structure with a large internal surface area and high surface area that behaves as a powerful adsorbent [1–5]. Activated carbon can be prepared by a physical or chemical activation process. The physical activation is performed in two steps, pyrolysis and activation in steam or carbon dioxide. In the chemical activation method, the precursor is impregnated with a chemical agent and heated for carbonization in an inert atmosphere. Various chemicals have been used as activating reagents mainly alkali (KOH, K_2CO_3, NaOH, and Na_2CO_3) [6–9], alkali earth metal salts ($AlCl_3$, $FeCl_3$, and $ZnCl_2$) [10–13] and some acids (H_3PO_4 and H_2SO_4) [14–18]. In order to minimize the cost of producing activated carbon, we must use low-cost precursors such as agricultural waste [6–8, 11, 12, 17–21]. Indeed, agricultural waste has an important potential for the preparation of adsorbents due to their abundance and renewability. Agricultural biomass waste is a renewable resource and constitutes an important potential that can be better managed and utilized to provide a broad range of useful chemicals and materials thus can eliminate pollution because of its disposal to the environment. Thereby, biomass wastes agricultural have a significant potential for the recovery of valuable materials. In this regard, activated carbons can be produced from almost all kinds of carbohydrate containing raw materials: grape stalks [22], coffee husks [23, 24], coconut shells [25, 26], corn cobs [7, 14], waste potato residue [21], wood of birch [27], walnut shell [28].

In this study, date palm rachis was used as a precursor material for activated carbon preparation. The date palm (*Phoenix dactylifera*) is a tropical and subtropical climate plant. In turn, the date palm nut becomes a residue, which has no better commercial value, only used by people to get some fuel value by burning. Lack of appropriate disposal methods causes environmental pollution. Consequently, the activated carbon preparation from the date palm is one method by which it can be utilized as useful material. For this reason, the objective of the present work is to prepare high surface area microporous activated carbon from date palm rachis using NaOH as the chemical activating reagent. The obtained carbon was characterized and used for the sorption of *o*-cresol from the aqueous medium in a continuous adsorption system. The experimental adsorption data were fitted to four breakthrough models, the Adams-Bohart, Thomas, Yoon-Nelson and Yan models.

2 Experimental

2.1 Chemical and Materials

Date palm rachises were collected from Gafsa farms (south of Tunisia). The material was treated as described previously [29]. Chemicals used in this study were of analytical grade. Hydrochloric acid (HCl, ACS reagent, 37%) and sodium hydroxide (BioXtra, $\geq$ 98%) were supplied by Sigma-Aldrich. Sodium carbonate (Na_2CO_3, $\geq$ 99%) was supplied by Fluka. Sodium bicarbonate ($NaHCO_3$, $\geq$ 99%) was supplied by Scharlau. NaCl (99%) was purchased from Merck KGaA, Darmstadt, Germany.

2.2 Preparation of Activated Carbon

The raw material (Date palm rachis) was placed in a horizontal stainless steel reactor and heated in a furnace at the rate of 5 °C min^{-1} under a nitrogen atmosphere for 2 h at 600 °C. After, the obtained biochar was mixed with different impregnation ratios (RAM = NaOH/biochar = 0.5, 1, 2, 3 or 4) and then dried in an oven at 105 °C for 12 h. Then, the samples undergo second pyrolysis at different temperature (500, 600 or 700 °C), and are kept at this temperature for (30, 60, 90 or 120 min) under an N_2 flow of 100 cm^3 min^{-1}. After cooling, the resulting activated carbon was washed with HCl (0.1 mol L^{-1}) followed by hot distilled water until neutralization (pH $\approx$ 7). Finally, AC was left to dry at 105 °C until a constant weight was obtained. The activated carbon was crushed and sieved; a fraction with 250–500 μm particle size was recuperated and stored in desiccators until use. The yields obtained for the preparation of AC were calculated according to Eq. 1:

$$\text{Yield}(\%) = \frac{w_f}{w_0} \times 100 \tag{1}$$

where w_f (g) is the mass of obtained activated carbon and w_0 (g) is the dry mass of raw material.

2.3 Characterization

The Chemical analysis was performed according to TAPPI standard methods. The holocellulose, Klason lignin, ash and α-cellulose contents were determined respectively using Wise et al. process [30], T222 om-06, T211 om-07 and T203 cm-99. Thermo gravimetric analysis was carried out using a thermo gravimetric analyzer (TA Instruments Q50, New Castle, Delaware). 10 mg of sample was heated from room temperature to 900 °C at a rate of 10 °C/min under nitrogen flow. The IR spectra were recorded in a Shimadzu 8400-S spectrometer in the range from 400 to 4000 cm^{-1}. The specific surface area (S_{BET}) and pore parameters of the carbon were obtained from nitrogen adsorption–desorption measurements at 77 K using a Micromeritics ASAP 2020 instrument. The scanning electron microscope (SEM) micrographs of the carbon were obtained using a ZEISS-ULTRA55 SEM microscope. The oxygen functional groups and the pH of the point of zero charge (pH_{ZCN}) were measured as previously described by Khadhri et al. [29].

2.4 Fixed-Bed Column Studied

Adsorption experiments in a fixed-bed column were performed in a 6 mm glass tube with a length of 60 mm coupled to a peristaltic pump for flow control. The studies were carried out with *o*-Cresol initial concentration of 216 mg L^{-1}, at room temperature, mass of activated carbon was 550 mg and at a flow rate of 1.5 and 3 mL min^{-1}. The samples were collected at the exit of the column at different intervals until the saturation of the bed. The concentration of *o*-cresol was determined using UV/Vis spectrophotometer.

To evaluate the adsorption in the fixed-bed column, breakthrough curves were studied by plotting the relative amount of *o*-cresol eluted (C_t/C_0) as a function of time. The operation of the column was stopped when the saturation state occurred (C_t/C_0 reached unity).

The exhaustion time (t_e) and breakthrough time (t_b) were taken when $C_t/C_0 =$ 0.95 and 0.05, respectively. The maximum capacity of the column (q_c (mg/g)), the total adsorbed quantity of o-Cresol retained in the column q_{tot} (mg), the total amount of *o*-cresol pumped into column m_{tot} (mg) and the removal of percentage R were determined using Eqs. 2, 3, 4 and 5, respectively.

$$q_c = \frac{q_{tot}}{m} \tag{2}$$

$$q_{tot} = \frac{Q\,A}{1000} = \frac{Q}{1000}\int_{t=0}^{t=t_{tot}} (C_0 - C_t)dt \tag{3}$$

$$m_{tot} = \frac{C_0\,Q\,t_{tot}}{1000} \tag{4}$$

$$R = \frac{q_{tot}}{m_{tot}} \times 100 \tag{5}$$

where C_0 and C_t (mg L^{-1}) are the inlet and outlet adsorbate concentrations, Q (mL min^{-1}) is the flow rate, m (g) is the mass of sorbent, and A is the area under the breakthrough curve.

Table 1 The mathematical models and their linear forms

Isotherm	Non-linear form	Linear form	Plot
Bohart-Adams [31, 32]	$\frac{C_t}{C_0} = \exp(k_{BA} C_0 t - \frac{k_{BA} N_0 H}{v})$	$\ln\left(\frac{C_t}{C_0}\right) = k_{BA} C_0 t - \frac{k_{BA} N_0 H}{v}$	$\ln\left(\frac{C_t}{C_0}\right)$ vs t
Yoon-Nelson [33]	$\frac{C_t}{C_0} = \frac{\exp(k_{YN} t - k_{YN} \tau)}{1+\exp(k_{YN} t - k_{YN} \tau)}$	$\ln\left(\frac{C_t}{C_0 - C_t}\right) = k_{YN} t - k_{YN} \tau$	$\ln\left(\frac{C_t}{C_0 - C_t}\right)$ vs t
Thomas [31]	$\frac{C_t}{C_0} = \frac{1}{1+\exp\left(\frac{k_{Th} q_0 m}{Q} - k_{Th} C_0 t\right)}$	$\ln\left(\frac{C_0}{C_t} - 1\right) = \frac{k_{Th} q_0 m}{Q} - k_{Th} C_0 t$	$\ln\left(\frac{C_0}{C_t} - 1\right)$ vs t
Yan [31, 34]	$\frac{C_t}{C_0} = 1 - \frac{1}{1+\left(\frac{C_0 Q t}{q_Y m}\right)^{a_Y}}$	$\ln\left(\frac{C_t}{C_0 - C_t}\right) = a_Y \left(ln\left(\frac{C_0 Q}{q_Y m}\right) + \ln(t)\right)$	$\ln\left(\frac{C_t}{C_0 - C_t}\right)vs\ln(t)$

k_{BA}–Bohart-Adams constant (L mg^{-1} min^{-1}), N_0–saturation concentration (mg L^{-1}), H–bed depth of the column (cm), v–linear flow rate (cm min^{-1}), k_{YN}–Yoon-Nelson kinetic constant (min^{-1}), τ–contact time required for 50% breakthrough of the adsorbate (min), k_{Th} Thomas kinetic constant (mL mg^{-1} min^{-1}), q_0–maximum capacity of the sorbent (mg g^{-1}), q_Y–Yan model maximum adsorption capacity (mg g^{-1}), and a_Y–constant of the Yan model

The experimental curves were analyzed using the Thomas, Bohart- Adams, Yan and Yoon-Nelson models (Table 1). The accuracy between experimental data and models were evaluated using the regression coefficient (R^2), overage relative error (*ARE*), residual root-mean-square error (*RMSE*) and nonlinear analysis chi-square test (χ^2).

3 Results and Discussion

3.1 *Characterization of Raw Material*

Date palm rachis was used as raw material for the production of activated carbon. Table 2 summarizes the chemical composition and elemental analysis of raw material.

Elemental analysis (Table 2) shows that the rachis of the date palm have relatively high levels of carbon and oxygen as well as other chemical elements with lower contents such as hydrogen, nitrogen and sulfur. This expresses the richness of the biomass in the carbonaceous matter. Taking into account the work from the literature, the holocellulose content of the date palm rachis (67.78%) is higher than that of *Posidonia oceanica* (61.8%) [37] and of certain annual plants such as *Retama raetam* (58.7%), *Pituranthos chloranthus* (61.8%) and *Nitraria retusa* (52%) [35]. The lignin content of the rachis (29.23%) is comparable to that of *Posidonia oceanica* (29.8%) [37] but it is higher than that of other perennial plant materials and agricultural residues. The ash content of the rachis was found to be 6.63%. It is higher than those

Table 2 Chemical and elemental composition of date palm rachis and some plants

	Ash (%)	Holo. (%)	Lig. (%)	Cell. (%)	C (%)	O (%)	H (%)	N (%)	S (%)
Date palm rachis	6.63	67.78	26.23	44.63	43.83	45.07	5.11	0.22	0.06
Astragalus armatus [35]	2.8	70.80	19.0	41.5					
Pituranthos chloranthus [35]	5	62	17.6	46.5					
Nitraria retusa [35]	6.2	52	26.3	41					
Retama raetam [35]	3.5	58.7	20.5	36					
Stipagrostis pungen [36]	4.6	71	12	44	42.89	42.21	5.86		
Posidonia oceanica [37]	12	61.8	29.8	40					
Opuntia ficus-indica [38]	5.5	64.5	4.8	53.6					

of olive wood (1.4%), *Retama raetam* (3.5%) and *Astragalus armatus* (2.8%) [35] but it is lower than that of *Posidonia oceanica* (12%). The contents of holocellulose and lignin are determining parameters in the carbonization yield and the properties of the carbon obtained [5, 39–41]. It appears that the chemical characteristics of the date palm rachis are interesting with respect to the experimental values presented in the literature. This makes it possible to consider that this biomass could be used as a precursor for the manufacture of activated carbon, potentially recoverable in various fields [5, 42].

The thermal decomposition behavior of date palm rachis (Fig. 1) shows three stages: the first stage at around 30–120 °C is related to the dehydration period and a weight loss of about 6% is due to moisture as well as some low molecular weight volatile compounds. The second weight loss stage, having a considerable weight drop (67%) occurs from 120 to 550 °C, due to the degradation of cellulose, hemicellulose and also a part of lignin. This stage is accompanied with a release of volatile compounds. The third mass loss observed for a temperature above 550 °C corresponds to the decomposition of the rest of the lignin up to 700 °C with a low mass loss of around 5%. Beyond this temperature, ash appears and a practically constant mass is obtained. Thus, we have noted that a temperature between 600 and 700 °C seems to be the most adopted for preparing activated carbon from the materials studied.

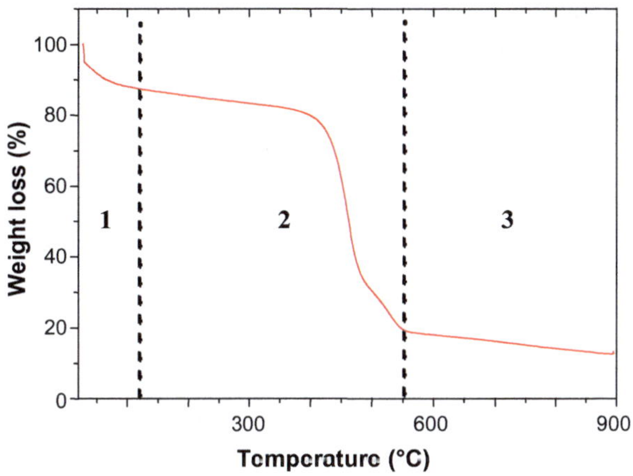

Fig. 1 TGA curve of date palm rachis

3.2 Preparation and Characterization of Activated Carbon

During the preparation of the carbons, we were interested in studying the effect of the pyrolysis temperature and the pyrolysis time on the yield of activated carbon for different activation ratios (0.5, 1, 2, 3 and 4). The yield of the obtained activated carbon reaches their maximum at a temperature equal to 600 °C for a RAM = 3 and a pyrolysis time of 2 h (Fig. 2). In the rest of the work, the activated carbon obtained under optimal conditions was characterized and used for the retention of o-cresol.

Activated carbon prepared under optimal conditions is mainly composed of 74.32% carbon, 14.79% oxygen, 2.30% hydrogen and 0.38% nitrogen. The basic groups were higher than the acidic groups (carboxylic, phenolic and lactonic), which is in agreement with the $pH_{ZCN} = 8$ (Table 3).

The FTIR spectrum shows an absorption band around 1630 cm^{-1} indicating the presence of carbonyl groups, or carboxylic groups (carboxylic acids or esters). In addition, an absorption band around 1030 cm^{-1} reflects the presence of groups (C-O-R). Note also the presence of a broad absorption band around 3400 cm^{-1} indicates in particular the presence of alcohols, phenols and carboxylic acids.

Scanning electron microscopy images of the surface and section of the activated carbon (Fig. 3) show a highly porous structure with open pores in the form of sharp, sharp edges. This proves the mesoporous structure of activated carbon with an average pore diameter of 51.36 Å and a specific surface area of 1108 $m^2\ g^{-1}$ (Table 3).

3.3 Continuous Adsorption

In this part, we studied the continuous adsorption of *o*-cresol on prepared activated carbon. The adsorption tests were carried out with a solution of *o*-cresol with an

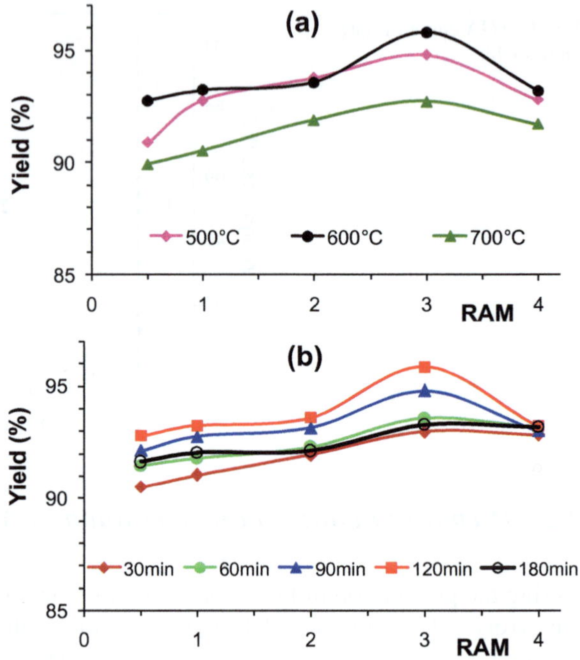

Fig. 2 Effect of temperature (**a**), impregnation ratio and duration of pyrolysis (**b**) on the yield of activated carbon

Table 3 Surface chemical analysis of activated carbon

Total pore volume (cm^3 g^{-1})	0.21
Average pore diameter (Å)	51.36
Specific surface S_{BET} (m^2 g^{-1})	1108
Carboxylic groups (mmol g^{-1})	0.172
Lactonic groups (mmol g^{-1})	0.183
Phenolic groups (mmol g^{-1})	0.087
Total basic groups (mmol g^{-1})	0.688
pH_{ZCN}	8

initial concentration of 216 mg L^{-1} for two volume flow rates of 1.5 and 3 mL min^{-1} (Table 4).

Curves obtained are shown in Fig. 4, and column parameters are listed in Table 4. The values of the maximum adsorption capacity of the column are 104.93 and 82.46 mg g^{-1} respectively for the flow rates of 1.5 and 3 mL min^{-1}. This decrease is attributed to the insufficient contact time of the adsorbate in the column to allow the solute to diffuse into the pores of the activated carbon. As a result, the breakthrough curve shifts to the origin (Fig. 4) and the column will quickly become saturated. However, a slight decrease in the percent elimination of o-Cresol is observed, which goes from 80 to 78% with an increase in flow rate from 1.5 to 3 mL min^{-1}.

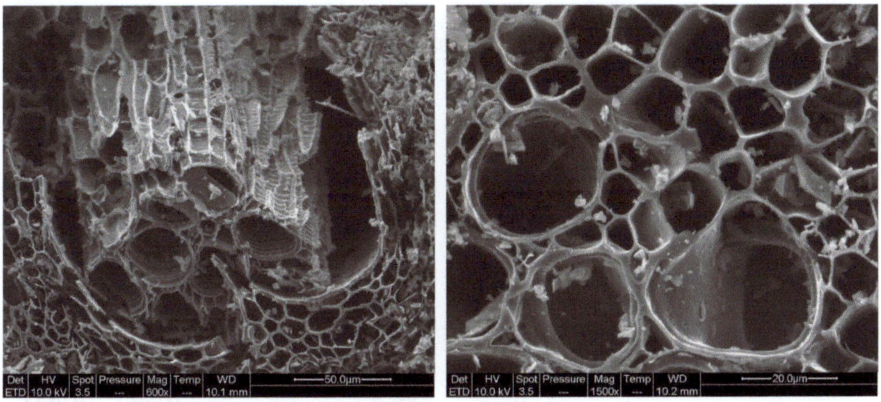

Fig. 3 SEM photographs of activated carbon derived from date palm rachis

Table 4 Fixed-bed column parameters for o-cresol adsorption

Flow rate (mL min^{-1})	t_b (min)	t_e (min)	q_c (mg g^{-1})	R (%)
1.5	130.40	221.95	104.93	80.25
3	47.10	89.63	82.46	78.08

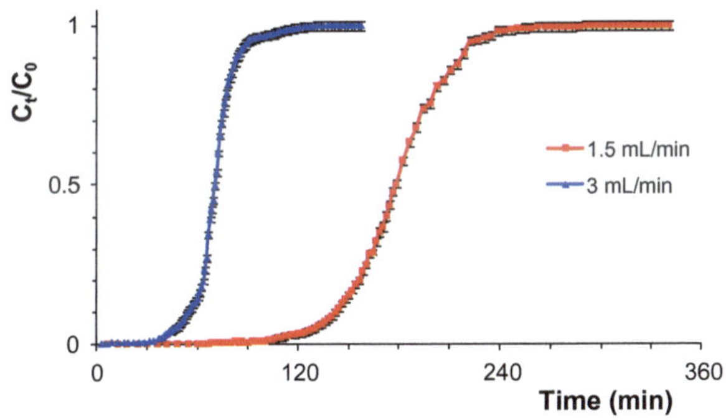

Fig. 4 Breakthrough curves for o-cresol adsorption onto activated carbon

The models of Thomas, Bohart-Adams, Yoon-Nelson and Yan were applied to describe the adsorption of o-Cresol. The results are shown in Table 5. The adsorption of o-Cresol follows the Thomas and Yoon-Nelson models well with regression coefficients (R^2) 0.9912 and 0.9853 respectively for the flow rates of 1.5 and 3 mL min^{-1} (Table 5).

The initial behavior of the breakthrough curves is perfectly described by the Bohart-Adams model with regression coefficients (R^2) greater than 0.98 (Table 5). This indicates that surface scattering is the limiting step in the adsorption process, confirming the validity of the Thomas and Yoon-Nelson models. This is also confirmed by the values of χ^2, RMSE and ARE. The validity of the Thomas model

Table 5 Bohart- Adams, Yoon-Nelson, Thomas and Yan model parameters for IC adsorption onto activated carbon

Flow rate (mL min^{-1})		1.5	3
Bohart-Adams	k_{AB} (L mg^{-1} min^{-1})	0.0002	0.0005
	N_0 (mg L^{-1})	104.60	90.16
	R^2	0.9872	0.9930
	χ^2	0.2172	0.1143
	ARE (%)	42	14
	RMSE	0.0309	0.0250
Yoon-Nelson	k_{YN} (min^{-1})	0.0602	0.1105
	τ exp. (min)	177.84	70.06
	τ cal. (min)	181.10	71.84
	R^2	0.9912	0.9853
	χ^2	0.1388	0.6658
	ARE (%)	33	16
	RMSE	0.0176	0.0582
Thomas	k_{Th} (mL mg^{-1} min^{-1})	0.2787	0.5116
	q_0 exp. (mg g^{-1})	104.93	82.46
	q_0 cal. (mg g^{-1})	106.68	84.65
	R^2	0.9912	0.9853
	χ^2	0.1388	0.6659
	ARE (%)	33	16
	RMSE	0.0176	0.0582
Yan	a_Y	6.8544	5.6577
	q_Y (mg g^{-1})	92.46	68.01
	R^2	0.8586	0.8214
	χ^2	5.1777	9.1361
	ARE (%)	131	116
	RMSE	0.1259	0.1685

is confirmed by the agreement between the experimental and calculated values of the column capacity. However, the close agreement between the experimental and calculated values of the breakthrough time (τ), suggests the validity of the Yoon-Nelson model (Table 5 and Fig. 5).

4 Conclusion

High surface area and mesoporous structure activated carbon have been prepared from date palm rachis by chemical activation using sodium hydroxide. The effects of the impregnation ratio and the pyrolysis temperature were studied. Results show that

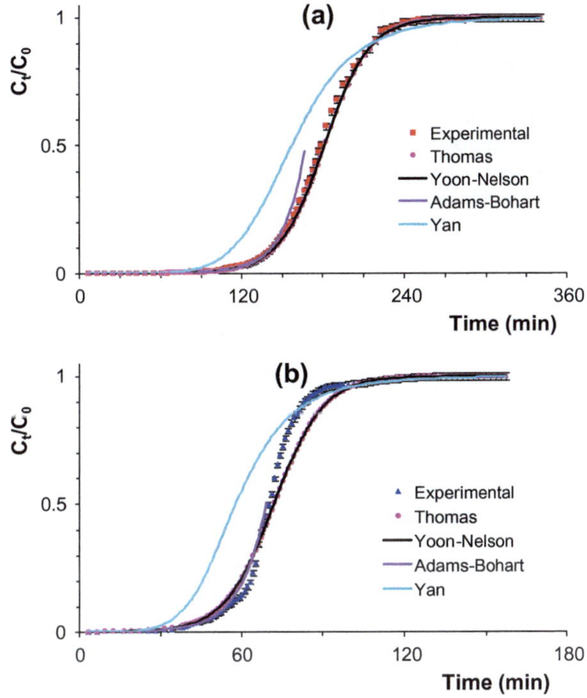

Fig. 5 Experimental and predicted breakthrough curves for the adsorption of *o*-cresol onto activated carbon at flow rates of: **a** 1.5 and **b** 3 mL min^{-1}

optimum conditions are a temperature of 600 °C, impregnation ratio of 3 and pyrolysis duration of 120 min. According to the Boehm titration, activated carbon surface is constituted mainly of basic groups with a $pH_{ZCN} = 8$. The prepared activated carbon was effectively used as adsorbents for the retention of o-cresol in continuous mode. The equilibrium data fit competently with Thomas and Yoon-Nelson models and the initial behavior of the breakthrough curves was well described by the Bohart-Adams model. These results indicated that date palm rachis could be used as a precursor to produce suitable adsorbents for o-cresol in wastewater.

Acknowledgements We greatly acknowledge the financial support of the Ministry of Higher Education and Scientific Research of Tunisia.

References

1. Suhas PJM, Carrott MMLR, Singh R, Singh LP, Chaudhary M (2017) An innovative approach to develop microporous activated carbons in oxidising atmosphere. J Clean Prod 156:549–555
2. Hassan AF, Abdel-Mohsen AM, Fouda MMG (2014) Comparative study of calcium alginate, activated carbon, and their composite beads on methylene blue adsorption. Carbohydr Polym 102:192–198

3. Kim TY, Cho SY, Kim SJ (2011) Adsorption equilibrium and kinetics of copper ions and phenol on to modified adsorbents. Adsorption 17:135–143
4. Shou J, Qiu M (2016) Adsorption kinetics of phenol in aqueous solution onto activated carbon from wheat straw lignin. Desalin Water Treat 57:3119–3124
5. Danish M, Khanday WA, Hashim R, Sulaiman NSB, Akhtar MN, Nizami M (2017) Application of optimized large surface area date stone (Phoenix dactylifera) activated carbon for rhodamin B removal from aqueous solution: Box-Behnken design approach. Ecotoxicol Environ Saf 139:280–290
6. Darweesh TM, Ahmed MJ (2017) Batch and fixed bed adsorption of levofloxacin on granular activated carbon from date (Phoenix dactylifera L.) stones by KOH chemical activation. Environ Toxicol Pharmacol 50:159–166
7. Vu MT, Chao H-P, Trinh TV, Le TT, Lin C-C, Tran HN (2018) Removal of ammonium from groundwater using NaOH-treated activated carbon derived from corncob wastes: batch and column experiments. J Cleaner Prod 180:560–570
8. Elhleli H, Mannai F, Ben Mosbah M, Khiari R, Moussaoui Y (2020) Biocarbon derived from Opuntia ficus indica for p-nitrophenol retention. Processes 8:1242
9. Pezoti O, Cazetta AL, Bedin KC, Souza LS, Martins AC, Silva TL, Júnior OOS, Visentainer JV, Almeida VC (2016) NaOH-activated carbon of high surface area produced from guava seeds as a high-efficiency adsorbent for amoxicillin removal: kinetic, isotherm and thermodynamic studies. Chem Eng J 288:778–788
10. Saygili H, Güzel F (2016) High surface area mesoporous activated carbon from tomato processing solid waste by zinc chloride activation: process optimization, characterization and dyes adsorption. J Cleaner Prod 113:995–1004
11. Siddique A, Nayak AK, Singh J (2020) Synthesis of FeCl3-activated carbon derived from waste Citrus limetta peels for removal of fluoride: an eco-friendly approach for the treatment of groundwater and bio-waste collectively. Groundw Sustain Dev 10:100339
12. Chen C, Mi S, Lao D, Shi P, Tong Z, Li Z, Hu H (2019) Single-step synthesis of eucalyptus sawdust magnetic activated carbon and its adsorption behavior for methylene blue. RSC Adv 9:22248–22262
13. Liu Y, Cheng H, He Y (2020) Application and mechanism of sludge-based activated carbon for phenol and cyanide removal from bio-treated effluent of coking wastewater. Processes 8:82
14. Sych NV, Trofymenko SI, Poddubnaya OI, Tsyba MM, Sapsay VI, Klymchuk DO, Puziy AM (2012) Porous structure and surface chemistry of phosphoric acid activated carbon from corncob. Appl Surf Sci 261:75–82
15. Lütke SF, Igansi AV, Pegoraro L, Dotto GL, Pinto LAA, Cadaval TRS (2019) Preparation of activated carbon from black wattle bark waste and its application for phenol adsorption. J Environ Chem Eng 7:103396
16. Liu QS, Zheng T, Wang P, Guo L (2010) Preparation and characterization of activated carbon from bamboo by microwave-induced phosphoric acid activation. Ind Crops Prod 31:233–238
17. Baysal M, Bilge K, Yilmaz B, Papila M, Yürüm Y (2018) Preparation of high surface area activated carbon from waste-biomass of sunflower piths: kinetics and equilibrium studies on the dye removal. J Environ Chem Eng 6:1702–1713
18. Han Q, Wang J, Goodman BA, Xie J, Liu Z (2020) High adsorption of methylene blue by activated carbon prepared from phosphoric acid treated eucalyptus residue. Powder Technol 366:239–248
19. Williams PT, Reed AR (2006) Development of activated carbon pore structure via physical and chemical activation of biomass fibre waste. Biomass Bioenerg 30:144–152
20. Ben Mosbah M, Mechi L, Khiari R, Moussaoui Y (2020) Current state of porous carbon for wastewater treatment. Processes 8:1651
21. Osman AI, Blewitt J, Abu-Dahrieh JK, Farrell C, Al-Muhtaseb AH, Harrison J, Rooney DW (2019) Production and characterisation of activated carbon and carbon nanotubes from potato peel waste and their application in heavy metal removal. Environ Sci Pollut Res 26:37228–37241

22. Ozdemir I, Ahin MS, Orhan R, Erdem M (2014) Preparation and characterization of activated carbon from grape stalk by zinc chloride activation. Fuel Process Technol 125:200–206
23. Getachew T, Hussen A, Rao VM (2015) Defluoridation of water by activated carbon prepared from banana (Musa paradisiaca) peel and coffee (Coffe arabica) husk. Int J Environ Sci Technol 12:1857–1866
24. Laksaci H, Khelifi A, Belhamdi B, Trari M (2017) Valorization of coffee grounds into activated carbon using physic—chemical activation by KOH/CO2. J Environ Chem Eng 5:5061–5066
25. Cazetta AL, Vargas AMM, Nogami EM, Kunita MH, Guilherme MR, Martins AC, Silva TL, Moraes JCG, Almeida VC (2011) NaOH-activated carbon of high surface area produced from coconut shell: Kinetics and equilibrium studies from the methylene blue adsorption. Chem Eng J 174:117–125
26. Talat M, Mohan S, Dixit V, Singh DK, Hasan SH, Srivastava ON (2018) Effective removal of fluoride from water by coconut husk activated carbon in fixed bed column: Experimental and breakthrough curves analysis, Groundwater. Sustain Dev 7:48–55
27. Bergna D, Varila T, Romar H, Lassi U (2018) Comparison of the properties of activated carbons produced in one-stage and two-stage processes, C 4:41
28. Li Z, Hanafy H, Zhang L, Sellaoui L, Netto MS, Oliveira MLS, Seliem MK, Dotto GL, Bonilla-Petriciolet A, Li Q (2020) Adsorption of congo red and methylene blue dyes on an ashitaba waste and a walnut shell-based activated carbon from aqueous solutions: experiments, characterization and physical interpretations. Chem Eng J 388:124263
29. Khadhri N, Saad MK, Ben Mosbaha M, Moussaoui Y (2019) Batch and continuous column adsorption of indigo carmine onto activate carbon derived from date palm petiole. J Environ Chem Eng 7:102775
30. Wise L, Murphy E, Addieco MAA (1946) Clorite holocellulose: its fractionation and bearing on summative wood analysis and on studies on the hemicelluloses. Paper Trade J. 122:35–43
31. Huang D, Wang G, Shi Z, Li Z, Kang F, Liu F (2017) Removal of hexavalent chromium in natural groundwater using activated carbon and cast iron combined system. J Clean Prod 165:667–676
32. Bohart GS, Adams EQ (1920) Some aspects of the behavior of charcoal with respect to chlorine. J Am Chem Soc 42:523–544
33. Sheng L, Zhang Y, Tang F, Liu S (2018) Mesoporous/microporous silica materials: preparation from natural sands and highly efficient fixed-bed adsorption of methylene blue in waste water. Micropor Mesopor Mater 257:9–18
34. Yan G, Viraraghavan T, Chen M (2001) A new model for heavy metal removal in a biosorption column. Adsorpt Sci Technol 19:25–43
35. Ferhi F, Das S, Elaloui E, Moussaoui Y, Yanez JG (2014) Chemical characterisation and suitability for papermaking applications studied on four species naturally growing in Tunisia. Ind Crops Prod 61:180–185
36. Ferhi F, Das S, Moussaoui Y, Elaloui E, Yanez JG (2014) Paper from stipagrostis pungens. Ind Crops Prod 59:109–114
37. Khiari R, Mhenni MF, Belgacem MN, Mauret E (2010) Chemical composition and pulping of date palm rachis and Posidonia oceanica–a comparison with other wood and non-wood fibre sources. Bioresour Technol 101:775–780
38. Mannai F, Ammar M, Yanez JG, Elaloui E, Moussaoui Y (2016) Cellulose fiber from Tunisian Barbary Fig “Opuntia ficus indica” for papermaking. Cellulose 23:2061–2072
39. Qin C, Chen Y, Gao J (2014) Manufacture and characterization of activated carbon from marigold straw (Tagetes erecta L) by H3PO4 chemical activation. Mater Lett 135:123–126
40. Mackay DM, Roberts PV (1982) The dependence of char and carbon yield on lignocellulosic precursor composition. Carbon 20:87–94
41. Elseify LA, Midani M, Shihata LA, El-Mously H (2019) Review on cellulosic fibers extracted from date palms (Phoenix Dactylifera L.) and their applications. Cellulose 26:2209–2232
42. Daoud M, Benturki O, Kecira Z, Girods P, Donnot A (2017) Removal of reactive dye (BEZAKTIV Red S-MAX) from aqueous solution by adsorption onto activated carbons prepared from date palm rachis and jujube stones. J Mol Liq 243:799–809

MIX
Papier aus verantwortungsvollen Quellen
Paper from responsible sources
FSC® C105338

If you have any concerns about our products,
you can contact us on
ProductSafety@springernature.com

In case Publisher is established outside the EU,
the EU authorized representative is:
Springer Nature Customer Service Center GmbH
Europaplatz 3, 69115 Heidelberg, Germany

Printed by Libri Plureos GmbH
in Hamburg, Germany